Keeping Your Kids Safe on the Internet:

A Parent's Guide to Technology

by Kerry Rego

Copyright © 2025 by Kerry Rego.

All rights reserved. No part of this publication may be reproduced, distributed or transmitted in any form or by any means, including photocopying, recording, or other electronic or mechanical methods, without the prior written permission of the publisher, except in the case of brief quotations embodied in critical reviews and certain other noncommercial uses permitted by copyright law. For permission requests, write to the publisher, addressed "Attention: Permissions Coordinator," at the address below.

Kerry Rego Consulting Books
P.O. Box 11463
Santa Rosa, CA 95406 USA
www.kerryregoconsulting.com

Book Layout ©2013 BookDesignTemplates.com
Cover & Graphic Design | Chelsea McKenna Design
Editing | Sunlight Editing
Photography | Michelle Feileacan Photography

Ordering Information:
Special discounts are available on quantity purchases by corporations, associations, and others. For details, contact Kerry Rego Consulting Books at the address above or book@kerryregoconsulting.com
Keeping Kids Safe on the Internet: A Parent's Guide to Technology / Kerry Rego. —1st ed.
ISBN 978-0-9906183-6-2

Table of Contents

Chapter 1 | How We Got Here

    Why Me?

    The School Speaking Tour Begins

    The Reason I Teach This Information

Chapter 2 | Everything Feels Out of Control

    Why Isn't This Being Taught In Schools?

    Storytime: Unexpected High School Lessons in 3 Acts

    Technical Education in Schools Does Exist

    Tech Champions

    Technology is not the Enemy

    Let's Talk Moral Panic

    Legal Age Limit

    Storytime: My 11-Year-Old is Going to Jail

Chapter 3 | The Danger Zones

Access Points

Monitoring in the Physical World

Monitoring in the Digital World

Chapter 4 | Staying Safe Online: Safety Action Steps

Inappropriate Content

What Kids Can Do: Safety Action Steps

What You Can Do: Safety Action Steps

When Your Kid is the One Posting – Let's Talk

Consequences

Critical Thinking & Media Literacy

Build Critical Thinking Skills: Safety Action Steps

Storytime: Recycled Disinformation

Critical Thinking in Real-Time

Storytime: It Happens to Me Too

Internet & Social Media Red Flags

Storytime: Shares That Hurt Like a Knife

Table of Contents

Chapter 1 | How We Got Here

    Why Me?

    The School Speaking Tour Begins

    The Reason I Teach This Information

Chapter 2 | Everything Feels Out of Control

    Why Isn't This Being Taught In Schools?

    Storytime: Unexpected High School Lessons in 3 Acts

    Technical Education in Schools Does Exist

    Tech Champions

    Technology is not the Enemy

    Let's Talk Moral Panic

    Legal Age Limit

    Storytime: My 11-Year-Old is Going to Jail

Chapter 3 | The Danger Zones

Access Points

Monitoring in the Physical World

Monitoring in the Digital World

Chapter 4 | Staying Safe Online: Safety Action Steps

Inappropriate Content

What Kids Can Do: Safety Action Steps

What You Can Do: Safety Action Steps

When Your Kid is the One Posting – Let's Talk

Consequences

Critical Thinking & Media Literacy

Build Critical Thinking Skills: Safety Action Steps

Storytime: Recycled Disinformation

Critical Thinking in Real-Time

Storytime: It Happens to Me Too

Internet & Social Media Red Flags

Storytime: Shares That Hurt Like a Knife

DON'Ts & DOs

Storytime: Overexposing Children

Sexual Exploitation

How to Protect Your Child: Safety Action Steps

Watch Out

Report Abuse

Storytime: Love Hijacked

Having "The Talk" About Porn

Teens Sharing Photos

Revenge Porn

Chapter 5 | Bullying

Bullying Facts

Signs Your Child is Being Bullied

Important Resources

Storytime: Sharing Resources with Community

Protect Your Child: Build a Family Safety Plan

Family Safety Plan for Parents

Family Safety Plan for Kids

Building Community and a Culture of Empathy

Chapter 6 | How to Regain Control

Where Your Power Lies

You Decide When the Time is Right

Storytime: My Family's Timeline

Learn About Technology

Increase Your Control: Safety Action Steps

Monitor Activity

Set Up Parental Controls

Storytime: The Flip Phone

Engage in Conversation

Chapter 7 | Create a Healthier Relationship with Tech

Build Healthy Life Habits

Decrease Tech in Your Life

Storytime: Gaggle is Tech-Free

Resources

    Book Recommendations

    Downloads

    Law Enforcement & Legal Assistance

    Learn About Tech

    Regain Control

    Tech Support & Parental Controls

Bibliography

Dedicated to

My beautiful Joy,
I'm so lucky to be your mom.
I'm grateful for our relationship
because I know not every mom and daughter
gets along as well as we do.

With all of my heart and all of my might
through every day and every night.

My sweet Dave,
I've never known love and support like that
which you give me.
My world hasn't looked the same
since you showed up
and I want it to always be lit by your heart.
Genesis 2:23

Dedicated to

My beautiful Joy,
I'm so lucky to be your mom.
I'm grateful for our relationship
because I know not every mom and daughter
gets along as well as we do.

With all of my heart and all of my might
through every day and every night.

My sweet Dave,
I've never known love and support like that
which you give me.
My world hasn't looked the same
since you showed up
and I want it to always be lit by your heart.
Genesis 2:23

CHAPTER 1

How We Got Here

I will never forget the mom who came up to me after a school safety presentation. She was on the verge of tears when she asked me if she could take her daughter's phone away. I asked if she was a minor – yes. I asked if the mom paid the bill – yes. I looked her dead in the eye and said, "You control this situation. You can absolutely take it away from her." She started sobbing with relief. I knew there were so many scared parents out there just like her.

In 2010, I started visiting campuses to talk to kids about staying safe online. I wanted to empower them to use critical thinking to identify misinformation and to understand the potential lifelong consequences of their actions. I wanted them to understand how to harness these tools to change the world.

The only other people doing this same work were members of law enforcement, but they were focused on crime and prevention, not technology. Tech wasn't being taught in the classroom in any consistent way, and that set off alarm bells for me. So I went everywhere I was asked: K-12 schools, community colleges, career programs, universities, afterschool programs, leadership summits – the works.

I didn't plan on this path of educating kids and their families about how to stay safe online – it chose me.

Why Me?

I'm an independent social media, digital marketing, and technology consultant working in the field since 2006. I help all kinds of businesses and organizations understand the how and why of using social media to promote their products and services. I create and update strategies, train staff and marketing teams, educate audiences as a speaker, give advice regarding compliance, and provide tech support on these rapidly changing tools.

At that time, there wasn't another vendor that provided social media education in my community. I became well-known for my ability to take this big, scary topic and turn it into something smaller, more manageable, and less terrifying. Word of mouth is powerful. People passed my name around, and I began to get asked to go into new spaces to speak to all kinds of audiences.

After years of presentations, school systems began recruiting me to teach the subject matter in curriculum offered courses. I've taught high school, community college, university, and professional development programs.

I am more than a little passionate about social media and technology. The field is constantly changing at warp speed. I feel responsible to my clients to have the latest information, so I put in a tremendous amount of time researching, reading, testing, and gathering public feedback to keep my expertise up to date.

The reason I'm writing this book now is because I'm a parent of a nineteen-year-old named Joy. Whew! We made it through. Raising my child in the age of social media was not easy. I want to guide parents who are still in the middle of navigating how to raise children safely in our technology-heavy world.

This book is:

- An active resource for parents.
- A comprehensive guide with further reading, data, and resources.
- Filled with personal stories of what worked for raising my kid in the era of social media.
- Everything that I believe parents could benefit from knowing about tech education.
- A message in a bottle containing words of support from one parent to another.

This book is not:

- A deep dive into all things tech.
- A solution to all your technology problems.
- A 100% fit for all families.
- Timeless – remember, this field changes at warp speed.

My goal is to give you back a sense of control and give you action steps to help your kids stay safe.

Did I do everything right regarding internet safety while raising my kid?

Of course not. But I've learned as much from the mistakes as from the wins. I know my experience can be of service to others. Just like we were sitting next to each other on the soccer field or at a kid's birthday party, I'll share what worked for us, knowledge I've gained as a professional in the tech space, and what I learned the hard way as I parented my little one.

I must acknowledge that after eighteen years of ferociously protecting my child's safety both online and off, it's rather strange to now be showing her off to the world. I avoided saying her name or showing her face on the internet for such a long time, and now she's embedded in this book, via story and video. You'll hear from her in Chapter 2.

The School Speaking Tour Begins

It kicked off in 2010 when a youth leadership camp asked me to share how young people could be productive with digital tools. I was so excited to share the possibilities of how tech can be used for good. I love giving aspirational presentations; I always walk out full of hope and energy. My goal is to inspire the attendees to feel empowered with their newfound knowledge.

Youth change the world, but adults don't give them enough credit for it. Look up any activist movement and you'll see young people on the front lines. I want to make sure the next generation of leaders knows how to use the tools they play with, like Instagram and TikTok, to make positive changes in the world around them.

In 2011, legal cases involving social media and schools began to appear in the news. That's when

I teamed up with Judith Delaney, an intellectual property attorney with a serious pedigree as representation for the major tech players. Together, we created a plan for campus administrators, complete with guidance around risk management for schools and internet safety education for students.

Together, we sent over forty-five registered letters to superintendents in our county. We did not get a single response. Upset at the radio silence, I spoke at my local county office of education public meeting. They looked at me with blank expressions and quickly moved onto the next agenda item. Not a single administrator nor elected official responded to our warnings of danger to come.

I can't express to you how angry this still makes me, to this day. As a parent and an educator, I consider it an absolute dereliction of duty.

In 2015, the adults started to get more concerned about the safety of their children on the internet. Calls started coming in from parent groups, principals, social services, and foster care programs asking me for a different type of presentation. Parents were scared, teachers were nervous, and kids didn't have anyone responsible who spoke their language. I went from teaching about being strategic online (i.e. making your dreams happen, creating a positive presence on the internet, etc.) to directly addressing the dangers our kids faced online. I began to understand the large gap in technology education. It dawned on me that I was one of the very few resources these groups had available. So I leaned in with all my might. I was offered a unique opportunity to help people and make a difference in their lives.

The Reason I Teach This Information

This wasn't in my career plan. Over fifteen years ago, I was enjoying my job as a consultant providing services to businesses and organizations, imparting my passion and advice about marketing and its constant changes. But once the panicked phone calls and emails started rolling in, they simply never stopped.

Truthfully, I don't love talking about how dangerous and scary technology can be, but it feels like a purpose that I've been given. I'm in the right place with the right knowledge, not only because of my job but also because as a parent, I was living it every day. I understand the overwhelming feeling that parents experience when they feel out of control because they don't understand the digital world in which their children live. When I think about the consequences and worst-case scenarios I've seen play out, there's no question that I'm going to do everything I can to help keep kids safe.

As a parent, I felt alone so much of the time when I was making choices for my daughter with all of this technology surrounding us. The parental controls were nothing like they are today. At the time, very few people were discussing how to protect their kids; I was making stricter choices than other parents, so I felt very alone. At the same time, I knew I was lucky because this was *my field*. But even with my knowledge, it was still confusing. Tools weren't set up with safety in mind, only profit. Every time I struggled with a setting, an app, or keeping up with it all, I knew that other parents were much worse off, adrift on the seas of technology with no life preserver. That's when I realized I was one of the lifeguards.

I wrote this book so you wouldn't feel alone. I want you to have as many resources as possible to navigate these scary waters.

Whether it's in person or through my writing, my goal is to reach as many parents, caregivers, schools, and educators who need me. I want to provide easy-to-understand information about a very confusing topic and bring relief to people who feel scared or lost.

I will come to your school or program in person, speak virtually, and record sessions, if you need. Because this field changes rapidly, I plan to continue to adapt the material as our needs change. I'll never stop doing what I do as long as the requests keep coming in.

overwhelming feeling that parents experience when they feel out of control because they don't understand the digital world in which their children live. When I think about the consequences and worst-case scenarios I've seen play out, there's no question that I'm going to do everything I can to help keep kids safe.

As a parent, I felt alone so much of the time when I was making choices for my daughter with all of this technology surrounding us. The parental controls were nothing like they are today. At the time, very few people were discussing how to protect their kids; I was making stricter choices than other parents, so I felt very alone. At the same time, I knew I was lucky because this was *my field*. But even with my knowledge, it was still confusing. Tools weren't set up with safety in mind, only profit. Every time I struggled with a setting, an app, or keeping up with it all, I knew that other parents were much worse off, adrift on the seas of technology with no life preserver. That's when I realized I was one of the lifeguards.

I wrote this book so you wouldn't feel alone. I want you to have as many resources as possible to navigate these scary waters.

Whether it's in person or through my writing, my goal is to reach as many parents, caregivers, schools, and educators who need me. I want to provide easy-to-understand information about a very confusing topic and bring relief to people who feel scared or lost.

I will come to your school or program in person, speak virtually, and record sessions, if you need. Because this field changes rapidly, I plan to continue to adapt the material as our needs change. I'll never stop doing what I do as long as the requests keep coming in.

I always remember the terrified parents who came up to me after I delivered a presentation in an auditorium or crowded band room. The pained looks on their faces, the frustration, the tears. When we talk, I see the beautiful relief that telegraphs through their bodies. I feel the connection that happens when they realize they have someone to talk to about their concerns. Those moments are beyond rewarding. Those moments are what drive me to keep doing this work. If I can help people in any way, and if I can leave that school safer, you bet I will.

CHAPTER 2

Everything Feels Out of Control

You probably picked up this book because you're scared for your kids and you feel overwhelmed when it comes to technology. Welcome to the club. But there's good news! It gets better.

It's natural to fear the unknown, but I want to break apart the belief that everything about technology and parenting is new; it's not as foreign as you may think. I've presented this information in person hundreds of times over the last fifteen years. I have a really good idea of how you'll react and feel as we go on this journey together, and I need to warn you that it's going to start out scary and then get worse before it gets better.

Hang on. I've got you.

I promise to bring you back out on the other side with resources, tactics, and tools to improve your situation. You *will* feel better by the end. I ride this rollercoaster every day, and I'm here to guide you through the scary parts. We'll do it together.

Need me right now?

Email me at k@kerryregoconsulting.com.

Or join my Parents & Families Group on Patreon

https://www.patreon.com/kregobiz.

In the coming chapters, expect that I'll get tough with you occasionally. This is love in action. A good friend won't just tell you what you want to hear, they'll tell you what *you need to hear*. I'm a parent too, so imagine us having this conversation while on the sidelines at the t-ball game or the grocery store. I'm also a teacher. That's a mom double whammy. I'm looking out for you and your babies just as I know you would for me and mine. This parenting club is a beautiful thing and it takes all of us working together to keep our families safe.

Here's my first tough statement: As a parent, it's your primary job to protect your children from danger. Yes, tech companies need to do a better job of protecting young people, but their priority is profit. We can never forget it. I remind people of this brutal reality because it shifts the conversation back to what we need to do as parents. We can't trust tech leaders to have our kid's mental health and safety – or ours – in mind. I can't say this more loudly:

It is no one's responsibility but our own to parent our children.

I understand that parents are exhausted, overwhelmed, and unclear as to what to do. Again, I'm a parent too, and even though my kid is past eighteen, my job as a parent isn't done. My job will never be done. I want to show you what I learned, what we tried, what the research shows, and how it worked in real time – not just theory. No one ever said parenting was easy – it's not. But improving the odds of your child's safety when it comes to the dangers of technology *can* be done. It's our job.

And to answer a question I know you have – my daughter didn't use social media (with permission) until she was fourteen. I worked *very* hard to make that happen, and still caught her sneaking devices, accounts, and generally trying to duck under the boundaries I set. I was able to keep her safe because I did everything I could, even when I didn't want to. Was it a breeze? Heck no, but it was worth every ounce of effort.

You can do it too. I'll hold your hand, and we'll get through it together.

(Scan for a video of me talking more about doing this together.)

Why Isn't This Being Taught in Schools?

I've been asking this question for over fifteen years. I do have an answer but you may not like it.

If you live outside of the United States (U.S.), your nation might have a robust technology program in schools to educate children eighteen and younger. In the U.S., where I live, we do not have any such plan or program. I live just two hours north of Silicon Valley, and even with proximity to the veritable tech center of the world, the whole thing is a disaster.

My child went through the public school system. I volunteered as a room mom and donated my time to teach Joy and her classmates in the computer lab. Since then, I've taught high school and college, and I'm a consulting vendor for K-12 institutions and universities. I haven't seen every school district in this vast country, but I've seen enough to know the multi-factor answer to this important question.

Change in the school system starts with legislators who write bills to prioritize, fund, and provide resources. Those resources then trickle down into the classrooms. I can't emphasize enough how massive a topic this is, but I've given you the flow of the factors that contribute below.

Consider the age and generational perspective of legislators. Consider their general lack of business acumen and the fact that the majority are lifelong politicians with ample societal and economic privilege. They don't understand the central place that technological skill sets have in today's economy.

This results in:

- Legislators not prioritizing technology education in schools.
- Lack of funding devoted to this subject matter.
- Lack of a strategic and consistent plan.
- Educational systems and requirements that are broken up by state, region, and school district in the U.S., which means it's difficult to have consensus on how to solve this problem.
- The rapid pace of innovation and student needs.
- Lack of time and professional education for teachers and administrators to champion change in their schools and districts.
- Lack of equity when it comes to high-speed internet access to ensure all kids can actually use the technology that's in place.

- The overemphasis on testing that results in other all-important life topics being left out of our children's curriculum.

It's not just technology and corporate regulation we're dealing with here. The lack of preparation for our children is a part of a much larger web. As far as what is not taught or emphasized in schools, it's not just technology education that gets minimized. Food science (healthy cooking and eating), financial literacy, art, music, physical exercise, and so many other important topics are not taught in schools because they are not viewed as priorities in our current model of education.

Parents have to face this unpleasant truth: likely, your child is not getting a complete education. Teachers do an amazing job with the little they're given, but let's be real: In no other industry is a professional required to pay out of their pocket to buy job-related supplies. Or to fundraise for materials that are essential to their job. Pilots don't buy fuel, engineers don't buy steel, and scientists don't buy lab equipment. We live in a system that does not equip educators for success, which means that our kids aren't equipped for success either. When I hear parents or employers lament, "Why don't they teach that in schools?" I understand their frustration. But we can't keep laying this responsibility (and blame) on teachers who are operating within a broken system. We must acknowledge that public education is missing the mark when it comes to preparing children for a tech-centered world.

In the past, parents educated their children at home on life lessons, while schools managed the book lessons. Something has shifted recently, and both sides of the equation are struggling to deliver on what they previously managed without noticeable gaps. The gaps in our children's education have become chasms.

Storytime:
Unexpected High School Lessons in 3 Acts

When I taught a high school business class, it was offered in a hospitality-focused core. Many of our students would graduate and seek jobs in that industry because the school is located in a top tourist destination: Sonoma County wine country. As a part of their graduation requirement, students had to obtain internships with either our school's hospitality partners or outside organizations they identified on their own.

When we gathered feedback from the businesses, we learned that the students didn't know how to speak to an adult on the phone. They did not understand that they needed to leave a voice message to get a return call so that they could schedule an interview.

In my high school classroom, many of my students were adults (or close to it) and they had never left a voicemail. Pause and think about that. I understand that it's a generational thing to not want to talk on the phone (I can't stand it myself), but making sure kids know how to talk to adults is important for their survival. What do you think will happen when they apply for a real job? *Their parents never taught them how to talk on the phone.*

My fellow teachers and I were stunned. We were forced to design and implement a new lesson to teach them this basic life skill. That lesson took time away from other important curricula. A lesson they should have learned at home.

When I realized these kids didn't know how to talk on the phone, I knew I could take the lesson one step further. I have a strong handshake; I've learned that in business, a lot of people never do. I had my students mingle like they were at a networking event, and I instructed them to shake hands with each other, make eye contact, and practice introducing themselves. There was lots of laughter and awkwardness – both normal reactions for people of all ages. I shook the hand of every student, gave them notes, practiced, and encouraged them so they could gain the confidence they needed.

Shaking hands is one of the very first things we do in life and business. It leaves a powerful first impression. It's the start of relationships and jobs. Long ago, many families taught their kids how to do this. That's no longer the case.

One day, I brought Dr. Frank Chong, the president from the local community college where I also teach, to speak to my class. He was more than happy to talk about the benefits of his school and answer their questions, especially because many of the students would be enrolling there after graduating from high school. My goal was to break down the mystery of college and put a friendly human face on it, especially a university president of color.

After he left, I had the students write thank-you notes in appreciation of his time. They each got a card I'd purchased, but I withheld the envelopes. I drew an envelope on the whiteboard at the front of the room and diagrammed how to address it, where the return address belonged, and where to place the stamp.

Do you know how many of my 30 students had never addressed an envelope?

Probably 90%.

I was raised writing thank-you notes; my grandma was a stickler for them. I've heard many complaints about how people don't write thank-you notes anymore, and for some reason, it's blamed on technology. I also hear how much people enjoy receiving handwritten notes and how rare they are today. We need to remember that this behavior (and gratitude, as a value) is learned in the home.

How did I know they needed the envelope lesson?

Years before, I had an intern who was a senior in college. I asked her to address, by hand, a stack of invitations for my first book release party. When she was done, I took a look at the stack. She had addressed them all center-aligned, shaped like a triangle. Normally, the name and information of the addressee are left-aligned, meaning all the first letters of each line are lined up on the left.

My intern was weeks from graduation and ready to go into the professional workplace and yet she didn't know how to address an envelope properly.

My point with these three stories is that teachers don't normally give lessons on things like using the telephone, shaking hands, or addressing envelopes. That's not in the curriculum because those are life lessons. Life lessons are best taught at home. Teachers cannot teach your children *everything* they need to know. I believe that school is a supplement to the education you give your child, not the other way around. I find this mindset helpful, too, when it comes to accepting the responsibility that it's our job as parents to keep our kids safe online.

Technical Education in Schools Does Exist

There is a robust technical education program called International Certification of Digital Literacy[1] (ICDL) that I wish the U.S. would adopt. Started in 1995, the European Computer Driving License (ECDL) certification program was developed through a task force and recommended by the European Commission High Level Group to be a Europe-wide certification scheme. In 1999 it was rebranded to the International Certification of Digital Literacy.

I learned about ICDL in 2011 when a representative of theirs wanted to work with me to push local adoption of the program. I was disappointed that our project didn't pan out, but I'm intrigued by their model. ICDL is the most recognized computer certification in the world. With it, students gain real-world skills that enable them to be more productive learners, employees, entrepreneurs, and citizens. ICDL can be implemented modularly at any time, making it easy for schools to fit into timetables.

ICDL has been at this for a long time and I think they've figured it out. They've successfully delivered certification programs to over 17 million candidates across more than 100 countries in the world. If there's a way to get it done, the ICDL model is the answer. If I were an education funder or lobbyist, this is where I'd be: making ICDL a required part of our children's education.

Tech Champions

If I learned anything about getting things done in schools, it's that every change needs a champion. This is one person who lights the spark that turns into a fire. The champion could be a parent or administrator, but in my experience, it was always a teacher.

Educators aren't in schools because they make a lot of money – most can barely get by. Most are highly educated and committed to their craft. They know what they do changes lives and impacts the future. Many of them teach technology as best they can, and embed it into their lessons and curriculum without a formal system supporting them. They know it's important to their students' success and that students love lessons that include technology. Teachers are doing the best with what they have available and I hate that it falls on them to address this giant gap.

Kids need to learn digital citizenship, literacy, critical thinking, research skills, and how to use technology. People of all ages need a basic understanding of tech to apply for jobs, perform at work, excel in school, access healthcare, interact with the government, and more ways that evolve each year. If your solution is to bin it, that's probably fear talking. Dictionary.com defines a person with this fear as follows:

> ***Technophobe*** *(noun):*
> *Someone who fears the effects of technological development on society and the environment, someone who is afraid of using technological devices, such as computers.*

If you identify as a technophobe, I get it. Technology unchecked is scary. Like any fear, we have to get to know it or it will rule our lives.

Technology is not the Enemy

Technology is a tool, and like any other tool, it is inert until used by a sentient being with motivation. Technology does what we ask it to do. In and of itself, tech isn't bad. It's neutral. If I hand you a pencil, you can write a love sonnet or a terrorist manifesto. If I give you a hammer, you can use it to build a breathtaking building or smash a window. It's not the technology, it's how we use it. It's the intent.

It's easy to say "ban it!" and never let your kids use technology, but there's a downside to that route. At some point, your kids will be accessing tech and the internet (if they aren't already). As a parent, you should know that a complete ban can result in your kids doing whatever they can to access the internet in a way that's not only unsupervised but also unsafe. When they encounter a problem, they aren't going to come to you for help. If the solution were as easy as turning it all off, we would've already done it by now. Life (and tech) is always more complicated than we wish it to be.

But all the bad things!

Technology is also not the only cause of mental health concerns, the increase in suicide rates, or the decrease in graduation rates. It's not the sole cause of – human trafficking, illicit drug use, lack of manners in society, or why your socks keep disappearing. It is a convenient boogeyman that we can pile all our concerns onto the back of when we want to find a reason for our problems. As a society, we struggle with the multi-layered causes, and we don't like to face the consequences of our choices.

I've seen much reporting that claimed technology definitely creates mental health problems. Those reports use data demonstrating correlation which is two things happening simultaneously, rather than causation which is when one thing causes another to occur.

The mental health of our children is important, and teens are reporting more sadness and poorer health than previous generations. If your first thought is that cell phones and social media have the biggest negative impact on their mental health, read a 2022 study of U.S. high schoolers by the Centers for Disease Control (CDC).[2] For me, two facts jump out of the study: the factor most associated with teen unhappiness is violent and emotional abuse by parents and household adults. The second is that abused and depressed teens are more likely to use social media, screens, and smartphones.

If we're serious about improving the quality of life for our kids, we need to look at every factor that impacts their health and we simply aren't doing that. We need to pay attention to the whole picture.

Let's Talk Moral Panic

Powerful social agents like the media and politicians use moral panic as a tactic to stoke public fears and concerns around an individual, group, or object. It's important for us to be wary of this manipulation because it can trigger irrational behavior in us as parents. (I talk about this more in the form of click bait in *Chapter 5*, especially around bullying.)

In Psychology Today, Dr. Scott A. Bonn explains how:

> *"Moral panic has been defined as a situation in which public fears and state interventions greatly exceed the objective threat posed to society by a particular individual or group who is/are claimed to be responsible for creating the threat in the first place."* (Psychology Today).[3]

Some examples of products, inventions, and trends that sparked moral panic – and were considered the end of civilization – include rock and roll, heavy metal, blues, jazz, rap, the Walkman, the sewing machine, magazines, comic books, and the flapper gal. You can dive into more examples via any internet search or by reading *Youth, Popular Culture, and Moral Panics: Penny Gaffs to Gangsta Rap, 1830-1996* by John Springhall.

One of my favorite visual examples of moral panic is a vintage image of riders on a crowded subway car in what looks to be the 1950s. Everyone in the picture is engrossed in their newspapers and the caption reads "Technology is making us antisocial." (Reddit)[4] We've always found a way to distract ourselves and fear change.

The reason I love working with technology is because I believe in the good and the possibilities that it brings. The American Academy of Pediatrics[5] published *Children and Adolescents and Digital Media,* which states that the benefits of technology "include early learning, exposure to new ideas and knowledge, increased opportunities for social contact and support, and new opportunities to access health promotion messages and information."

But I'm not going to insult you by pretending there aren't serious problems with the constant barrage of all things digital. That same paper also explores risks of technology, which "include negative health effects on sleep, attention, and learning; a higher incidence of obesity and depression; exposure to inaccurate, inappropriate, or unsafe content and contacts; and compromised privacy and confidentiality."

My point is this: technology didn't create our problems, but it does have a hand in exacerbating what we face. Our most common solution tends to be introducing legislation, also known as techno-legal solutionism, that aims to "save our youth" and solve everything that's wrong in our kids' lives. Technology legislation is difficult to pass, and even if it did pass, it wouldn't solve our larger issues (Angel and boyd).[6] We'd go looking for another boogeyman to blame for what ails us.

In addition to the academic paper cited above, I want to share another quote by technology researcher danah boyd (lowercase is her choice) from a piece she wrote on LinkedIn called *KOSA*

Isn't Designed To Help Kids (boyd).[7] It's about the Kids Online Safety Act that was proposed legislation in 118th Congress (2024-2025) but did not pass. It will be introduced in the 119th Congressional (2025-2026) session. She is a Microsoft researcher and a professor at Georgetown and Cornell Universities, among a much deeper tech pedigree. Her book *It's Complicated: The Social Lives of Networked Teens,* is in my recommended reading section (at the end of this book). I've been following boyd's educated and critical remarks on this industry for at least a decade. She breaks down why technology is part of a much larger puzzle:

> *"People keep telling me that it's clearly technology because the rise in depression, anxiety, and suicidality tracks temporally alongside the development of social media and cell phones. It also tracks alongside the rise in awareness about climate change. And*

the emergence of an opioid epidemic. And the increase in school shootings. And the rising levels of student debt. And so many pressures that young people have increasingly faced for the last 25 years. None of these tell the whole story. All of these play a role in what young people are going through. And yet, studies are commissioned to focus on one factor alone: technology."

I mentioned already that I've spoken at a youth leadership camp almost every summer since 2010. These thirty teenagers are smart, motivated, creative, and interesting – they're typical teenagers. I ask them to tell me their biggest concerns about the world today and they always come up with an intelligent, emotionally mature, and nuanced list. I write them on a whiteboard and then they vote on their top five concerns.

Their top worries during the summer of 2024 were in order:

1. Racial injustice
2. Drug abuse
3. Climate change
4. Violence against children
5. Reproductive rights

In 2023, their top worries were:

1. Climate change
2. Mental health
3. Reproductive rights
4. Teacher shortages
5. Healthcare costs

Other concerns on their long list included: book banning, the spread of hate and violence, anti-LGBTQ+ laws, gun violence, inflation, political polarization, geo-political unrest, capitalism, human trafficking, ableism, and several more complex social issues. These kids can see more of the big picture than we realize, but they're hopeful about making change.

I'm realizing now that bullying and cell phones rarely, if ever, make it onto their list of serious issues that need to be solved. I checked my records.

Which is to say that our kids are smart, empathetic, and more aware of the world around them than we were, in part because they're exposed to more of it. Technology is a big part of their lives but it isn't the center as much as we think it is. In fact, I'm seeing a swing of the pendulum back the other way, perhaps in response to growing up possibly overexposed on their parents' social media accounts. They see how glued we adults are to our devices, and many are finding solace in a more "retro" relationship with tech. They have big concerns about the world we're leaving them and simply banning them from using Snapchat isn't going to make all the bad things go away.

We have to find a way to strike a balance with tech.

Legal Age Limit

Did you know that thirteen is the minimum age for social media sites?

Legally these technology companies can't allow your child to use their site until that age. They risk big fines otherwise. Each social media service requires a birthdate to allow the user access or to create a profile. Since it's self-reported, it's easy to lie about. It also provides plausible deniability on the part of the platform.

Thirteen is the age set by Congress in the Children's Online Privacy Protection Act (COPPA)[8] passed back in 1998. COPPA is one of the few federal privacy laws in the United States. The internet has changed drastically since it was put in place which was well before smartphones and social media were even on the radar.

COPPA states that apps, websites, and online tools must:

- Provide notice and get parental consent before collecting information on children under thirteen.
- Have a clear and comprehensive privacy policy.
- Keep the information they collect from kids confidential and secure.

Just because it's the law doesn't mean thirteen is the right time for your child to use these tools. I knew when my kid approached thirteen that it wasn't safe enough for me to let her play on what amounts to an autobahn of dangerous activities. Many legislators and medical professionals are trying to get the minimum age raised to fifteen or even sixteen. This is a hot area of legislation and activity around the world.

The U.S. Surgeon General, Dr. Vivek Murthy (2021-2025), says kids under fourteen should not use social media (CNN).[9] "The skewed, and often distorted, environment of social media often does a disservice to many of those children." It's difficult enough for adults to navigate the media manipulation and misinformation that we encounter online, and yet we're surprised when our kids can't manage it safely.

Then why are kids using it?

One of few things are happening:

- Parents don't know about the age limit.
- Parents don't realize it's an actual law rather than a suggestion.
- Parents allow their children access regardless of their age.
- Kids are doing it without permission.

My dad used to say to me when I was begging him to let me do something he considered unwise, "If everyone jumped off the Golden Gate Bridge, would you?" and I hated it because I knew he was right. Being a tough parent, or the child of a tough parent, is not fun. In hindsight, I'm grateful for the parenting I received and I've heard from my daughter that I didn't ruin her life.

Proof! I've got proof! Scan the QR code to hear directly from Joy.

If the thought goes through your head that all the other kids are doing it, or that there's no way to stop your kids from accessing unsafe sites, or that they'll be terribly left out of social situations, then I have to stop you right now.

Don't cop out!

You *do* have control. You likely bought that device (even if you didn't, they're still a minor), you control access to the internet, and no, your child will not melt if they aren't able to use TikTok.

You're the boss. Don't let your children drive this car and convince you to let them use social media. They will crash themselves into a brick wall. You are the legally responsible driver here so take the wheel.

As a parent, you have control over:

- Their devices.
- Their internet access.
- The time they spend on both.

I can hear your objection. *Who has 100% control over everything their kid does?*

You know the answer – no one does. But that shouldn't stop us from setting boundaries and holding strong. "No" is a complete sentence and you need to practice saying it firmly. We'll come back to this.

I've heard reports of kids as young as nine having access to social media. I find that frightening. There's a massive difference between a nine-year-old and a thirteen-year-old navigating social. This is absolutely within our jurisdiction and I regularly have to remind parents just how much power we have when kids are that young. Parenting has always been hard and all this tech brings a heck of a challenge to our already difficult job. If necessary, find a parenting buddy who shares similar boundaries and values. Together, support each other through this difficult process and share what is and isn't working for your family. You don't have to navigate this landscape alone.

Storytime: My 11-Year-Old is Going to Jail

When my daughter was eleven, I was reviewing her device activity and I discovered she had a secret Twitter account. As I peeled back the hood and investigated her activity, she saw on my face that she was in trouble. She did her best to convince me that it wasn't a big deal. Then, when she figured out she was cooked, she tried to grab the phone away from me. I could see her growing panic as I looked through her account to see who she was following and who was following her. I grew worried seeing strange adults on her follower list. I was shocked that many of her friends (at eleven!) were using Twitter because, at the time, it was the preferred social platform of adult actors and sex workers advertising their services.

Note: Today Twitter is called X and it's much worse with misogynists, Nazi sympathizers, white supremacists, right-wing content creators, and crypto bros all sitting right alongside the porn and sex work.

While I navigated her account, I talked to her about how dangerous the site was for someone her age. I then told her that the law stated anyone under the age of thirteen cannot use social media, and that I was going to delete her profile. She misunderstood me and thought she was the one breaking the law. Her expression grew horrified because she thought she was going to go to jail. I did my best not to laugh out loud at her interpretation but kept a poker face. I corrected her and explained the law more clearly but I let her keep some of that fear. I am a mom, after all.

The worst part of that situation?

After I deleted her account, I knew that dozens of her fifth-grade friends were still on Twitter with no lifeguard, swimming in a pool with sharks. Those kids were completely unprotected. I wished I knew their parents so I could share with them what I'd discovered. The truth is that I knew even if I told them, some wouldn't do anything to protect their children. Not because they didn't care about their kids, necessarily; many wouldn't believe how dangerous a place it could be.

That was not the last time my kid got under the safety barriers I had established. Each time she did, I gave her the chance to be honest about her behavior. She'd lie about what she'd done, and then I'd show her what I'd discovered. We would talk about why it was unsafe for her, and then I'd delete the account. Yes, she'd get upset with me but anytime you discipline your kids that happens. I'd confiscate her devices for a time, tighten her security settings, and when she'd get her device privileges back, I'd increase the frequency of my monitoring.

We did this dance again and again. It was frustrating to have to keep doing it, but that's what I signed up to do. She didn't suffer undue harm from being left out of her friend's group chat for a little while, though she was angry with me. But she caught up when the time was right for her.

I want to leave you with two thoughts.

1. As the minimum age for social media use, thirteen isn't a suggestion. You aren't sneaking them into a PG-13 movie here. You're exposing them to very real threats on the internet. Threats that can be mentally and emotionally damaging as well as potentially deadly and dangerous for your kids.

2. You have more control than you think. Just like a muscle, you need to practice exercising that control. Kids don't like boundaries, but they often express

gratitude for them when they're older. Be kind yet firm. Keep telling them that you're setting those boundaries because it's your job to protect them. They need you to be the strong parent on this, not the fun parent. The stakes are too high.

CHAPTER 3

The Danger Zones

Since the moment they were born, or otherwise entered our lives, we've had to cope with the fact that our kids inhabit a dangerous world. This is a fact regardless of whether they were born in 2025 or 1925. When it comes to technology, most of us don't know what we're dealing with, understand the concepts, or even know how to pronounce the words.

You might say to yourself, "I didn't grow up with this stuff, so I don't get it." I hear this every time I talk to parents and caregivers. We are in a special spot in history where technology changes faster than our collective knowledge can adapt. Most of the adults around during our developmental years couldn't teach us what we needed to know to help us raise our kids today. They couldn't have anticipated the lessons needed for living in a microchip-centered world. The challenges facing parents have changed, and so must we.

Here's some tough love

It doesn't matter that you didn't grow up with it. This isn't about what you wish or want, it's about what you *need to do*. It's time to dust off your learning cap and *get learning*. Their lives depend on it. Get started by asking questions, finding resources (like you found this book, good for you!), and never stop learning because the tools will never stop changing. Don't be afraid to look stupid because this fear is what stops many of us from growing. It's time for your desire to protect your children to outrank everything else. (See the *Resources* section for additional learning options.)

Remember that your kids have been handed powerful digital devices with practically no lessons, education, or guardrails. They aren't smarter than you or wired to understand tech from birth. They're simply willing to try, to learn. They're figuring it out via experimentation, YouTube videos, and bad advice from friends. If they can figure it out, so can you. (See *Chapter 6 / How to Regain Control*).

The purpose of this section is for you to have a clearer understanding of the technology that you may not notice or even consider dangerous. You need to know about all the tools that are monitoring, watching, and tracking both you and your children. This list does not include every existing technology, just the ones applicable to this conversation. When it comes to keeping our kids safe in this digital world, here's where to start.

Access Points

When thinking about where your child accesses the internet and social media, you've likely thought about their phone or tablet.

Consider these additional entry points:

- Home computers
- School computers
- Friend's computers and devices
- Desktops, laptops, tablets
- Cell phones
- Mp3 players
- e-Readers
    - Ex. Nook, Kindle

- Game consoles and handheld gaming devices
    - Ex. Nintendo, PlayStation, Xbox
- Smart home appliances
    - Ex. Televisions, refrigerators, and home assistants such as Siri and Alexa

Everyone in the family needs to recognize that when we talk about safety on the internet and our devices, the conversation should touch on or reference *all* of the above.

Not everyone realizes that there are also variations of the web. For our purposes, I'm going to break up the internet into three neighborhoods:

1. Websites
    - A set of related web pages located under single domain name
    - Ex. Google, Netflix, Wikipedia

2. Social Media
    - Technology that allows the sharing of ideas and posting of content, and encourages social interaction
    - Ex. Facebook, Instagram, YouTube
3. Messaging Apps
    - One-to-one or private group messages that are designed to live inside apps, but may be available on the general internet
    - Ex. Messenger, Snapchat, WhatsApp

All of these can be accessed on a computer or a device through a web browser such as Chrome, Safari, Firefox, or Microsoft Edge. All of them have apps that can be downloaded onto devices. Many can also be used on smart appliances like a TV or refrigerator (weird, but true). You need to know that these three neighborhoods are neither synonymous nor identical. They have different features and functions, and your kids will use them very differently.

The point here?

You need to control and monitor all of those access points.

As explored earlier, your child's school does not have the capability to teach what you'll read here. If you don't talk to your kids about these risks, it's possible that no one else will. Just like we teach them to cross the street, we must point out the dangers our kids might encounter online.

Let's be realistic in what we can achieve.
- *Will you be an expert at all of these?* No.
- *Will you even remember everything listed?* Also, no.
- *Do I put into practice everything I've given you here?* No, I don't.

I'm not superhuman, and I'm not perfect. But I learn more each day, and I apply what's appropriate for my family and my needs. Give yourself grace.

Monitoring in the Physical World

These lists will *feel overwhelming* because they are. Take your time reading through, breathe, and come back if the information makes you feel anxious. That's to be expected.

Tech in the physical world can include:

- Aerial photography
    - Drones, helicopters, etc.

- Airport and transportation security checkpoints
- Banks and ATMs
- Building security cameras
- Google Street View cars fitted with cameras
    - Note: this populates Google Maps and other mapping services
- Government and law enforcement property
- Smart home devices
    - Ex. Doorbells, voice assistants, TVs, refrigerators, robot vacuums, etc.
    - Note: You can also encounter these as a guest in other people's homes and in rentals like AirBnB
- Sports and entertainment stadium surveillance
- Street and traffic cameras

Pretty standard stuff that you're already familiar with, yes? Seeing it all lined up might make you feel uncomfortable. It makes me uncomfortable, too.

A Thought on Surveillance

You may be thinking, "It's not a big deal to be monitored if you're not doing anything wrong," but have you thought about what happens with your data when for-profit companies or malicious "bad guys" access and exploit it? The potential and risk of misuse by a variety of approved (and unapproved) parties is endless.

This topic is much bigger than I want to get into with this book but please know that the places we are being monitored are increasing exponentially, security is lax (Techspot),[1] legislation is miles behind, capitalization of our information is extremely profitable (Proton),[2] companies lie to our faces (Forbes)[3] (The Record),[4] and opportunities for abuse are growing. Assume the worst when it comes to how our data is being used.

Monitoring in the Digital World

Let's move on to the newer and less visible tracking devices that watch us and record behavior. I'll go into greater detail with this list so you understand the capabilities and implications of each.

Less visible tracking devices:

- Algorithms
- App downloads
- Apps for hiding photos
- Beacons
- Bluetooth

- Computer cameras
- Deepfakes (Artificial Intelligence or AI)
- DNA and biometrics
- Ephemeral or disappearing messages
- Form fills
- Global Positioning System (GPS)
- Smart home devices (this is on both lists for a reason)
- Link clicks
- Personal Identifiable Information (PII)
- Social media followers
- Wearables
- Wi-Fi

Algorithms

It's important to talk about algorithms. It's a word that people throw around regularly, but rarely understand what it means or its implications.

Algorithms are a set of processes, or rules, to be followed in a problem-solving operation. I like to think of them as recipes that computers follow to get to a desired outcome. Technology platforms process millions and billions of pieces of content every day – images, posts, etc. These algorithms are programmed to make decisions about what is or isn't useful, interesting, dangerous, illegal, or appropriate for a given audience.

Each user has a profile (or custom recipe) that is comprised of their demographics such as the user's age, location, gender, and interests (based on their activity). The tech platforms use the recipes to make sure you're a happy and engaged user and getting the service or content you expect. It's not necessarily placed *on* you, it's designed *for* you so that you can enjoy your time there – and stay on site.

Many people talk about algorithms as if they're magic or a mysterious black box but they aren't. Some algorithms are proprietary and like the secret Coca-Cola recipe, they can be locked in a safe. But leave it up to ethical or white hat hackers, and we'll eventually learn a tremendous amount about how they're designed.

The content served to you in your social feeds and search engines is a result of your preferences, behavior, and the agenda of the platform itself, or the country and government in which you live.

They don't always get it right, but algorithms are mostly a reflection of our behavior. You interact with dog content, you'll get more dogs. Interact with negativity or hate speech, you'll get more of that. When it comes to the latter, hate speech content descends into a pipeline where the user is served more and more inflammatory and deadly information via recommendations. (Gonzalez v. Google)[5]

Tech Tip: You can monitor your child's YouTube video history. This is an important place to pay attention to their behavior! (See *Resources / Tech Support*). They may never tell you that they're watching videos that are radicalizing their thoughts, ideals, or behavior, but this is where much of that activity takes place. If you check in periodically on what they're watching, this will give you a lot of information that will help you keep them safe.

YouTube Kids isn't much safer. Many of the videos are miscategorized, and an unmonitored child is likely to end up on a video that seems designed for small children (based on the title or thumbnail) when in fact it is seriously problematic.

Never let your child use these tools unsupervised, even in "roped off" digital spaces.

Back to algorithms. Understanding how they work is key. The activity of the user is a major determinant of what content algorithms show you. I've encountered people who don't understand the concept, while some dismiss the explanation and choose to blame the companies as if that's the only factor at play. Actions have consequences. Consider the amount of time on site, what you read, accounts you follow, or content you click on, and you'll notice how quickly the content changes based on your behavior.

It's ironic that the song "Every Breath You Take" (1983) by The Police is the song with the most radio plays in history (Far Out Magazine)[6] because it's so prescient of our current reality. The lyrics demonstrate classic stalking behavior and that's exactly what today's tech does. As long as there are corporations making money off of our online activity to sell data to advertisers, this will be the case. Every link you click on, every account you follow, and every engagement you leave adds up. We must teach our children to be mindful in their tech use and that their choices follow them all over the web and can potentially impact their lives offline.

App Downloads

Programs or applications (apps) will ask for a variety of permissions upon install or first use. The app goes into setup mode and needs permissions so that all the features work seamlessly. Common permissions for access: Your contacts, location, microphone, and camera. They may share some or all of this information with other companies for profit or when subpoenaed. You won't be told how much of your info is shared and or with what entity.

You don't have to allow all the requests on the list, but then the app may not work correctly or even at all. If an app asks for something I'm not comfortable with and that causes it not to work, I delete the app. I'm not willing to compromise my safety or boundaries, or those of my child, all for an app. Remember that you can always review your permissions for any app. Just keep in mind that revoking permissions might mean the app ceases to function correctly, but you can take another crack at controls at any point.

Tech Tip: Review the permissions your apps have by going into your main settings and adjusting or restricting access.

Android devices tend to struggle with malicious apps more than Apple devices (NordVPN).[7] This is because Apple inspects all apps that go into their store more carefully, while the Google Play marketplace has an open approach that increases opportunities for bad guys to take root. Apple isn't impervious to malware, but it's much less likely to happen.

Some apps (and the companies that make them) straight-up lie about what they do. TikTok used to fall into this category, but after increased scrutiny on their practices, they've mostly cleaned up their act. The most egregious violator is the shopping site Temu (ClassAction).[8] In both 2023 and 2024, Temu was the #1 most downloaded iOS app in the U.S. (Apple)[9] and also the #1 Android download in the U.S. for 2024 (Business of Apps, data by Appfigures).[10] If you have this app, or your kid does, delete it now!

Delete Temu Now!

Android devices can download apps that exist outside the Google Play store which increases their vulnerability to downloads designed to do harm. These malicious apps, also known as malware, can steal your contact list and passwords, make your device send costly text messages, install adware that forces you to view ads and pop-ups, force load web pages, or download apps without your permission.

To be safe, you can change the setting on your child's phone or device so that it prevents them from downloading apps without your permission. (I talk more about this in *Chapter 6 | How to Regain Control*.)

Apps for Hiding Photos

Your kids could have secrets hiding in plain sight. Photo vault apps look like innocent applications yet their true purpose is to hide photos and data. The players in this space change on a regular basis but you can search for "apps to hide photos on my phone" or "secret storage apps" to get lists of the ones to know about.

Here are a few: Hidden Vault, Locker, Vaulty, and Secret Photos KYMS. Many of these apps disguise themselves as a calculator. Take a look at the calculator on your phone, then look for sneaky apps so that you can tell the difference between the default design and a "special" calculator.

If you find one of these on your child's device, it's time to shut the train down. There's no innocent reason any person, child or adult, would have one of these apps if they aren't hiding something serious. If I were in that situation, I'd confiscate the device, research if it has a "self-destruct" option before telling my child that they lose all of their devices until the app is opened for inspection. I need to know what's going on, plain and simple. I'll figure out what the next step is once I see what's inside.

Beacons

I loved the 2002 Tom Cruise movie *Minority Report*. It's science fiction but today it comes across more like a documentary. There's a scene where he's on the run and his eyes are scanned as he's moving. A Gap store calls out to him by name and suggests he repurchase a product he bought in the past. It was scary then, but it's now a real feature.

Beacons are digital signals sent out in a location, often a store, that registers your device and sends you a coupon or marketing message. You may have experienced this when you're at Target via their Cartwheel app. That app will tell you what's on sale and the exact location of those products in the store. They use Bluetooth beacons built into their lightbulbs to achieve this effect.

Don't like it?

Turn off your Bluetooth and leave it off.

There are more malicious ways this is being used but I don't really want to give you nightmares.

Bluetooth

Bluetooth is a wireless technology that exchanges data between devices within a short distance. This technology is an open standard, meaning that anybody can use it without a license.

It's convenient to have, and there are lots of uses, pros, and cons. I will focus on the security risks.

Leaving Bluetooth on all the time leaves devices vulnerable to bad actors who can:

- Steal content off a device;
- Steal your identity;
- Send spam and phishing messages that may contain malware;
- Spy on your activity;
- Impersonate you;
- Use the device to engage in a severe distributed denial of service attack (DDoS) or shut down the device;
- and broadcast audio through the speakers.

Tips to protect your devices from Bluetooth attacks (Norton):[11]

- Keep your operating system up-to-date;
- Turn off discoverability in Bluetooth settings;
- Avoid sharing sensitive information this way;

- Don't connect with just anyone over Bluetooth;
- Turn it off when not using;
- Turn it off after connecting with others in a secure location;
- and delete unused connections.

Computer Cameras

There are cases of cameras being turned on without consent or knowledge and without the indicator light turning on (The Guardian).[12] The U.S. government can and will monitor you via your computer camera if they deem it necessary. It's built into The Patriot Act using "Sneak and Peek" warrants (ACLU).[13] This means that you can be monitored or surveilled in violation of your Fourth Amendment rights.

How do you protect your child?

Put tape or a sticky note over your camera. Even Meta CEO Mark Zuckerberg does it (ABC News).[14]

I don't recommend using a plastic guard that has a small door you can slide back and forth over the camera. For years, I used a sticky note because they have gentle glue that won't damage your camera. I "upgraded" when I received a promotional gift from a local tech company. I loved it! Then my MacBook screen cracked as a result. It cost me $700 to replace. It wasn't until later that I saw the warning on Apple's website not to use plastic camera guards (face palm). Just use a sticky note trimmed down to a small size. Easy to replace and no extra cost.

Deepfakes (Artificial Intelligence or AI)
This isn't a new concept, but the influx of consumer-facing and accessible artificial intelligence tools (think ChatGPT and other AI

image generators) is one of my biggest areas of concern.

A deepfake is a video of a person in which their appearance has been digitally altered so that they look like someone else (Oxford Dictionary).[15] Audio deepfakes are also becoming increasingly common. These deepfakes are often created with malicious intent or to embarrass the individuals involved.

Digital media manipulation has been around since the beginning of printed media. Most recently, it's been done using Photoshop, often to put a celebrity's head on an adult entertainer's body. But now we've entered a whole new realm. We're starting to see young people use this tool against each other and this trend will continue to grow (Law & Crime)[16] (LA Times).[17]

If you ever share images and videos of your children, the existence of easy-to-produce deepfakes should give you pause. Your content is the seed for artificially created images you don't want out there. I don't recommend putting out any more media of your children than you have to. Adjust your account and social media settings to higher security levels and think carefully about what you share. Pause before posting. Ask yourself, "Why am I posting this right now?" An alternate way to share is to create a family group chat via text or a private messaging tool. This is a safer way to share and keep loved ones abreast of happy events, milestones, and growth.

Here's how I handled it: I wasn't concerned about sharing photos of Joy until she was around five when I began doing safety talks at schools. I went into my Facebook and Instagram accounts and deleted almost every photo of her. I switched to sharing media of her with family only via text and email. When I did share on the web, it was only a partial shot of her face. I rarely wrote her name in the post, and called her "my kid" instead. I found an app that offered a blurring option (free app Photo Editor+) for times when I needed to share a photo of her or another minor. Many people now only share photos that block or blur their children's faces. Again, Mark Zuckerberg does the same thing (CNN).[18]

DNA & Biometrics

This is my nightmare.

Biometrics are biological measurements used for identification. The most common types are finger and palm prints, facial recognition, voice recognition, iris recognition, and vein pattern recognition. Up until 2023, I'd only heard of it in commercial settings, such as accessing your place of work, but now Amazon encourages you to scan your palm at the grocery store and the gym. I say, heck no!

DNA risks are most commonly seen in the genetic testing space where you can learn about your ancestry. There are so many reasons sharing your DNA is a bad idea but here are the top five risks of sharing your DNA according to CNBC.[19]

Top risks of sharing your DNA:

- Hacking;
- Companies profiting off of your information;
- Few laws to protect your rights;
- Ability for law enforcement to access it;
- and the company changing its privacy statement to your detriment.

Want more reasons not to share your DNA?

See the *Bibliography* for a 2022 investigative report from Consumer Reports.[20]

In 2023, one of the most well-known companies in this space, 23andMe, was hacked. People with specific genetic traits, such as Ashkenazi Jews, were targeted, and the data was for sale on the dark web. Even worse, 23andMe *didn't tell their users* (Bleeping Computer).[21] Investigative journalists broke the story. Without their reporting, who knows if 23andMe would have ever revealed this data breach.

Then, in March of 2025, 23andMe filed for bankruptcy, and "federal law does little to secure genetic information given over to a private company" (NPR).[22] California Attorney General Rob Bonta issued a press release with instructions on how to delete your data if you are a user (California Department of Justice),[23] and I recommend you protect what you can.

See why it's my nightmare?

We've reached the point in our dystopian timeline where we need to talk to our kids about protecting their bodies, including protecting their biometric information. Update your body safety talk with your kids because this field has endless implications for their safety and future. You don't have to anticipate every possible risk; just teach them to be mindful and think twice before giving away this type of data.

Ephemeral or Disappearing Messages

Ephemeral messages are commonly known as disappearing or self-destructing messages. Once sent, they disappear after a set time, leaving no trace of the conversation. They aren't dangerous in themselves, but they are a fantastic way for either party to hide bad behavior, grooming, or illegal activities.

We first saw this in social media on Snapchat, and it was their key differentiator, the reason people used it. It was so popular that every other tech platform copied the feature (all tech companies take each other's ideas), and now you can find them as Stories on all Meta apps (such as Facebook, Instagram, Messenger, and WhatsApp) and elsewhere.

Be aware that any service that offers a disappearing message feature requires higher security settings – and more boundaries on your part, as a parent. More monitoring too, and potentially a ban all together; this was the route I chose for my kid.

Form Fills

Most websites have some sort of form that you can fill-in, such as subscribing to a newsletter, providing your information to get a coupon, or filling out a contact form. Most of the risk is around credit card and payment form fills, but here are some red flags to look for.

Red flags to look for:

- Forms sent via email link;
- Sites that don't use the https:// protocol in their URLs;

- Sites that don't have the lock icon in or near the browser bar;
- and sites lacking a privacy policy or those that don't talk about how they collect, use, store, and secure data.

This is a big one to talk to your family about because it's common that kids will go to a website and use a variety of email addresses to get benefits, coupons, or free trials. They are pros at the form fill. Talk to them about the risks.

GPS

We've been using Global Positioning System (GPS) technology for many years. GPS tracks a physical item's exact and real-time position using satellite technology. You might also see it used in apps or services called location-based services (LBS) or geolocation that use GPS to provide useful information about our surroundings such as where to find restaurants or gas stations, or play Pokémon Go.

We see GPS in our smartphones, our cars, our logistics and transportation apps, and so many other ways.

Ways you may be using GPS include:

- Navigation via tools like Google Maps;
- Mapping and adding specific locations into a customized trip or route;
- Recreation such as hiking, geocaching, and recording bike or run routes;
- Tracking belongings via Apple AirTag or Tile, and
- Traffic avoidance embedded in navigation tools.

Transparency with your kids about tracking and respecting privacy is key. Have conversations about what it means, the purpose of it, and what the expected behaviors are. This is most relevant if your kids use apps such as Snapchat

(SnapMaps), Apple devices (FindMyPhone), or Find My Friends to track each other.

Concerns about tracking include the lack of control over personal information, minimal awareness of who has access to that information once permission is given, abuse of the technology by stalkers, and the lack of industry privacy standards. There are many instances of companies asking for access to our physical locations to track our movements via apps and devices, yet not doing their due diligence in protecting that information from the prying eyes of employees or other vendors. It's shocking how little this data is protected.

Be mindful that your kids can be tracked by good guys and bad guys. The data can be sold, and there's very little legislation protecting us as consumers, both adults and children.

Link Clicks

Back to "Every Breath You Take," you know that every single link you click on is tracked. If you didn't before now, I'm sorry to have to break it to you. There's a record of the first time as well as the tenth time you've clicked, where you were when you saw the link, the path you traveled through the website, how long you were there, what you did, and so much more. It's all meticulously documented, used in advertising, and in other more nefarious ways.

But what about private or incognito windows?

Those aren't actually private: the marketing of what they do hasn't been clearly communicated to us. In some instances, we've been outright lied to (CNN).[24]

> *"Incognito or private mode will keep your local browsing private, but it won't stop your internet service provider (ISP),*

> school, or employer from seeing where you've been online. In fact, your ISP has access to all your browsing activity pretty much no matter what you do." (Mozilla).[25]

Your breadcrumb trail of clicks is a large determinant of what kind of content is served by the algorithm, via both organically posted content and paid ads. You can see what I mean if you click on an ad on Instagram. Almost instantaneously, you will be dogged by that same product or something similar for weeks or months afterwards everywhere you go on the web.

Behavioral retargeting is the action that advertisers use to show you promoted content that's related to the link you clicked on. This can happen in search engines, on social media, in a variety of other apps, and even in physical ads. Retargeting is used on people of all ages and is one of the ways your kids come in contact with content that's not appropriate or safe for them to consume.

Your kids might not know the connection between clicking on links and unwanted information, so it's up to the adults to talk with their kids about the consequences of digital actions.

Personally Identifiable Information (PII)

According to the U.S. Department of Labor,[26] personally identifiable information is "defined as information: (i) that directly identifies an individual (e.g., name, address, social security number or other identifying number or code, telephone number, email address, etc.) or (ii) by which an agency intends to identify specific individuals in conjunction with other data elements, i.e., indirect identification. (These data elements may include a combination of gender, race, birth date, geographic indicator, and other descriptors)."

You wouldn't think that kids are at risk of identity theft, but according to Experian, 25% of children will have their identities stolen before they turn eighteen. They are fifty-one times more likely to have their identity stolen than adults (Forbes).[27] It can be stolen by a thief, but it can also be stolen by a family member.

The lesson for your kids: Don't share this type of information with anyone or any service without asking for parental permission first. This is where close monitoring by parents is important. It also means that we never get to figuratively walk away from our kids when they're on a device. Yes, I know it's exhausting.

Smart Home Devices & Digital Assistants

This is a big category with a lot of different names like home devices, smart appliances, internet of things (IoT), and digital voice assistants. Technology that may be in your home include: smart refrigerators, robot vacuum cleaners, digital door locks with cameras, security cameras, baby monitors, pet cameras, smart TVs, Apple HomeKits, Google Assistants, and Amazon Alexa products.

All of these items have two elements in common:
1. They record your behavior via camera or audio recording.
2. They send that data back to their manufacturer.

It's remarkable how much information they gather, how little security there is around these devices, and how frequently rights are violated using this data. The security risk is high for you, your family, your neighborhood, and anyone in or near your home who doesn't consent to being recorded.

I have a hard and fast rule: I don't pay to be surveilled in my own home. If you feel comfortable giving those companies literal maps of your private space (robot vacuum), recordings of your comings and goings (doorbell cameras), and the ability to listen to and record everything you do, that's your prerogative. I've seen hundreds of instances of violations of privacy, and I just won't risk it. My family's safety is too important.

Social Media Followers

I grew up in the theater and, due to those years of exposure, I'm very comfortable on a stage with a bright light that obscures my ability to see who's looking at me. I never forget I'm being watched by people I can't see. Many internet users block out this reality. Maybe they can't truly grasp it. The truth is that hundreds, thousands, or millions of eyes are watching your every move when you broadcast via the internet. You can't see them and you don't know who they are, but that doesn't change the fact that your content is available for everyone to witness.

Our followers are often not who they say they are. It might be a kid looking for anonymity on a site they aren't supposed to use, a fake account that's harvesting data, a bot trying to incite negative responses from other users, or a predator befriending your child while using the profile picture of another person as a mask. You must act as if the people who follow you, the ones who aren't your real-life friends, are not there to be nice or good. I know being suspicious sucks, but in this instance, it can save lives and reputations.

Now you must teach your kids this survival tactic. Every single account that follows your child, even when your child's account is set to private or restricted, must be confirmed. If that follower is an actual real-life friend, wonderful. Occasionally, our friends will be hacked or impersonated, so don't follow them twice without confirming who they *actually* are. If it isn't someone you can confirm is real, then assume they're a pretender of some kind.

This is the new and improved Stranger Danger lesson our parents gave us. Everyone is a stranger unless we can prove otherwise.

Wearables

This technology is any kind of device that can be worn on your body. Wearables include: Smartwatches, fitness trackers, VR headsets, smart jewelry, Bluetooth headsets, and web-enabled glasses or augmented reality (AR) eyewear.

While great for health tracking, wearables pose similar risks as GPS does. One potential issue I see with wearables is when kids violate the privacy of others by wearing AR glasses into the gym locker room, and no one knows they're being recorded.

Wi-Fi

Are you willing to admit that you're not sure what Wi-Fi is, even though you probably use it every day?

It's one of those mysteries, similar to how electricity arrives in our homes and workplaces. Most people couldn't explain it. It had been a while since I learned it so I had to look it up to share with you, because it's important to understand.

Wi-Fi, also called WLAN, is a wireless network that has at least one antenna connected to the internet and wireless communication devices (such as laptops, computers, cell phones, etc.). A Wi-Fi network uses pulsed electromagnetic frequencies (EMFs).

There have been many scientific studies that research the biochemical effects that Wi-Fi has on the human body. Here's a quick synopsis and citing of studies from News-Medical.[28]

Biochemical effects of Wi-Fi:

- Oxidative damage to cellular macromolecules such as proteins, lipids, and DNA;
- Sperm count, motility, and DNA integrity;
- Lower testosterone levels;
- Elevated cell death;
- Reduced production and secretion of estrogen and progesterone in women, which reduces reproductive efficacy and impairs fertility;
- Chromosomal mutations in fetal development;
- Development of anxiety-like behavior;
- Disrupted learning and memory;
- Sleep deprivation; and
- Fatigue related to reduced melatonin secretion.

That said, the studies conflict with each other. There's a tremendous amount of information, but not a clear consensus. You might have forgotten that Wi-Fi has only been around a short time (since 2004) when it was first packaged into our handheld devices. Twenty years of data is not enough for long-term studies.

You need to at least be aware that Wi-Fi use – especially for your developing children – isn't without risks. How much of a danger it truly is to us is inconclusive.

Let's talk about a different kind of risk. Using unfamiliar Wi-Fi hotspots can expose you and your kids to other dangers. If you've ever experienced a difficult time trying to locate the correct hotspot you want to join because there are multiple names possible, there might be a Wi-Phishing trap waiting to catch you. This form of phishing uses a tactic called a "man-in-the-middle attack." MITM is an unauthorized interception of network traffic where the bad guy uses a poorly secured public or private Wi-Fi hotspot to intercept a transmission between the user and the website.

The best way to avoid a shady MITM hotspot (they're usually free) is to use Wi-Fi that requires a password. Whether you're at the local coffeeshop, on campus, at a hotel, or airport, you'll want one that requires you to log-in via the internet with a password provided by staff.

The setting "ask permission before connecting" is a wise choice on all devices so that your technology doesn't automatically join a network with a strong connection that has a bad guy lying in wait. All it takes is someone with malicious intent, the right equipment, and the know-how to spoof a Wi-Fi network's name. They can scoop your data, and who knows what they'll do with it.

If you must use a public Wi-Fi network, avoid doing sensitive tasks such as paying bills, accessing bank accounts, or making purchases where you have to pull out a credit card.

Are these things your kids are going to be doing in public?

Maybe not for a few years, but they grow up quickly. It's smart to teach them this life lesson now. They need to know what digital activities are safe for public spaces and those that should be done on a known and secure network.

You might have a free hotspot on your phone or other device; check the services available in your plan or package. When I'm preparing for domestic travel, I go to my local library and borrow a free hotspot unit. Not all libraries offer this, but do a little research on what technology services your local library offers. Mine has laptops, 3D printing, hotspots, podcasting studios, still and video cameras, tech lessons, and a whole lot more. These are your tax dollars at work.

Oh my gosh, that was a lot. We made it through this level. You deserve a cookie and a break.

CHAPTER 4

Staying Safe Online: Safety Action Steps

I want to acknowledge the energy you're putting into this. Good job taking the time to learn more and protect your children better when they're online. It's easy to leave it up to someone else. You're doing the hard thing, and I'm proud of you.

When I'm hired by administrators to teach online safety, this chapter is the meat of what I get into. Each community or campus has a pressing need or behavioral concern, and their requests are specific and emotionally charged.

Recent presentations have included:

- Fifth graders quoting misogynist influencers from TikTok, derogatory and violent language;
- An afterschool club wanting to make sure at-risk teens understood the dangers of social media and how it could impact their future; and
- A high school that wanted students to gain tools and guidance for handling difficult conversations in person, rather than lashing out at each other online.

Not only do I address the immediate concern, but I also try to cover some foundational groundwork that my audiences are missing. It needs to fit within the constraints of a short presentation (one to two hours) that is also interesting and useful. If it were up to me, I'd cover a lot more about monitoring tools online and off, hence me going all in on *Chapter 3 | The Danger Zones*. You're getting the bonus edition here, which covers all the ground rules about staying safe online.

Adults need to review their rules regularly with their children, as do schools with their students. Going forward, you'll see a symbol (see below) to signal there's a download that goes with the content. You can print them out to put on your fridge or somewhere central to help you weave the tips into your everyday conversations. The full list is available at the end of the book. It's time to set your rules of the road.

It's Up To Us

Corporations will never have your family's best interest at heart. Remember that they have a tight focus, and the safety of your kids is not on that list, no matter what they say. Actions speak louder than words, and consistently, the actions of technology corporations are at direct odds with our health and safety.

They focus on:

- Selling products;
- Making a profit for their shareholders;
- Tracking user behavior to serve and sell advertising.

It's our responsibility to raise our kids; technology cannot and will not ever do that. Let's get started.

Inappropriate Content

The internet has some regulations that can vary by location, but no matter where you live, there are nowhere near enough safeguards to suit the needs of minors.

Dangerous and damaging content your child will encounter on the web includes:

- Inappropriate or illegal behavior;
- Pornography;
- Threats of violence;
- Excessive violence;
- Hate speech; and
- Underage drinking, drug use, and sales.

What Kids Can Do

Think of this like the coaching you gave them about not talking to strangers. Give them action steps they need to know when they come across inappropriate content. With the fourth action step, do your best as the adult here not to blow up at them when it happens. Remember: You want to build trust, not break it.

Safety Action Steps, Kids:

1. Turn off the screen.
2. Hit the back button.
3. Close the app or game.

4. Tell a trusted adult if you feel scared or even if you think you'll get in trouble.

What You Can Do

You are one of the most trusted people in their world. Your reaction can have a tremendous impact on their views and emotions. Like it's an emergency drill, you need to mentally prepare yourself for how you will react. You need to practice your response so that you can be the very best version of yourself when it does happen. Think ahead of time what emotions may arise in this situation, and what you can do to self-regulate when the time comes.

It's a common reaction to feel fear or anger in this situation. But recognize those emotions and navigate them mindfully. Remember that it used to be that kids had to try hard to find inappropriate content on the web, but that's no longer the case. Much of the time, it's served to them on a platter or it's an innocent mistake. Don't beat yourself up if you don't do everything perfectly. You may have already gone through this, and you didn't have any resources. It's okay, you're here now, and you're learning. Just keep their best interest at heart. My loving advice to you is to be softer. If there's a time to be gentle with them, it's now.

Safety Action Steps, Parents:

1. Don't frighten or yell at them.
2. Listen attentively and stay calm.

3. Tell them it's not their fault, and check in on their emotions.
4. Tell them you're always there to help them, and that you're glad they told you.
5. Answer their questions as completely as you can, with age-appropriate responses.
6. Help them report the content to the website or the app where they saw it and to the Cyber Tipline (https://report.cybertip.org).[1]
7. Review security settings on the app or device to enhance protections.

When Your Kid is the One Posting – Let's Talk Consequences

If you were a hard-headed kid like I was, you might have needed to make the mistake to learn the lesson. Hearing a lecture from an adult about theoretical situations isn't always the most effective teacher. Experiencing consequences is sometimes the only way. Here are the consequences you can share with your child to help them get to the lesson faster.

They need to understand that much of what we do in the world is tracked digitally. They must know how they are being recorded in a variety of ways that can impact their life. Our systems are getting smarter using artificial intelligence (AI), facial recognition tools, and advanced search functions, meaning that their behavior matters and that many of these machines don't forget.

Behavior such as posting inappropriate content and online bullying leaves evidence and eyewitnesses. Disappearing messages don't disappear completely (kids are noticeably shocked by this.) The evidence (can include posts, comments, time stamps, recordings, etc.)

gathered from our devices and the web has long-term impacts.

Long-term impacts (across various aspects) include:

- Social Groups
    - There is a social cost. Their friends may choose to end relationships.
    - Entire grade levels or schools have been known to ostracize a student for their actions or views.
    - Clubs and extracurricular groups they are a part of will also be impacted.

- Family & Personal Physical Safety
    - Parents, siblings, and extended family members can experience impacts and retaliation that affect their jobs, reputation, and physical safety.

- Real-time reporting of locations can lead to theft, assault, and other types of danger to yourself and your family.

- Schools & Universities
    - School transfers may be necessary.
    - School activities and sports participation may be revoked if they're found to be breaking school rules.
    - Colleges perform social media background checks, and content may affect college acceptance.
    - Scholarships may hinge on behavior.
    - College acceptance offers can be rescinded if inappropriate online behavior is discovered, especially at private and religious universities that have stricter behavioral standards.

- Dating
    - Being rude, disrespectful, misogynistic, or lying about your past is easily discoverable.
    - Communities have sprung up online to call out awful behavior, sharing information and resources about individuals who engage in a variety of dangerous activities online and off.

- Job Search
    - Social media background checks are often standard procedure.
    - Job offers can be rescinded after the discovery of bad behavior or behavior that violates the organization's values.

- Financial Aid (Credit Cards, Loans, Government Assistance, etc.)

- You are being monitored, and will lose assistance if they deem your online behavior to be violating your request for aid.
- Even after receiving aid, if you are suspected of violating the conditions of the aid, they can take you to court and retroactively demand payments.

- Law Enforcement & Military
    - Poor choices can lead to arrest and/or conviction.
    - Social media background checks are standard procedure to get a job or to join the military.
    - Records of activities from the previously mentioned monitoring can and will be used against you in court.

When you talk to your kids about expectations of behavior, what is inappropriate, and what the consequences are, you'll want to run through some hypothetical scenarios to test their understanding. Ask questions on how they'll handle a situation, how they would respond, and what their next step would be. Get their brain working rather than allowing them to glaze over and tune out when this topic comes up.

Doing this regularly is necessary; repetition matters. As a parent and in school presentations, I would bring up recent stories in the press. I'd tell them about a recent episode, explain how it was damaging or illegal, talk about the consequences, and ask how they'd handle it if the same thing happened to them or someone in their friend group. Ask questions and get them to engage with the reality of the situation. They need to feel it and use critical thinking.

Critical Thinking & Media Literacy

Kids need our guidance and lots of practice to build their critical thinking skills. They need to learn these skills in multiple places – at school, at home, and in every setting that involves making the right choices. Imagine critical thinking as a muscle that needs a lot of exercise with a trainer (that's you).

A study by the News Literacy Project[2] found that 80% of teens who use social media say they see content about conspiracy theories at least once a week, and 81% believe at least one story. More than half failed to accurately identify an article as an ad or opinion piece. This is basic media literacy. It's not surprising because only 39% reported having had any media literacy training.

When it comes to a lot of things in their lives, kids get a bad rap. Adults spend a lot of time talking about how "young people today...." and then shake their "in my day" stick at them. Any opportunity I have to defend their intelligence, innocence, and unique predicament of this time in history, I will. Remember: adults are the responsible parties here.

Today's media landscape is uniquely challenging. Even adults need to unlearn bad habits, especially because they are in the position of being guides and teachers to young people online and off. I work with adults more than children and I'm shocked at the lack of critical thinking that I witness daily.

All it takes is to open Facebook (home of the older tech user), and you'll spot a post shared by your little Aunt Annie or Grandpa Joe to notice that they're rarely engaged in any sort of evaluation of information or fact-checking. They also seem to have no ability to identify AI-generated content that is false (also called AI slop). Anytime I see an erroneous declaration or rumor, I find a vetted source and gently share it with the person in the comment section. Do they take it down? It's hard to tell. Do they ever respond with an "oops"? Rarely.

According to a study published in the National Library of Medicine,[3] people over the age of sixty-five were more likely to share false information on the web than any other group. We spread "fake news" because we aren't paying attention and we move too quickly (Nieman Lab).[4] I am also guilty of this. But after several instances where I should've slowed down and checked my facts before spreading falsehoods, I've learned it doesn't take long at all to search to see if the story is true. Slowing down is an important step you'll teach your kids as we move through this chapter of Safety Action Items.

My favorite critical thinking question is, "*Who benefits from me reading or viewing this?*" which I learned from a fantastic resource created by Global Digital Citizen. It's a critical thinking exercise filled with questions designed to help us evaluate any form of information we're presented with.

The handout is no longer available from its source, but it's one of the most useful tools I've ever seen for development of these skills. Print it out and stick it in a high traffic area in your house. I memorized four-five questions on the sheet based on the "*Who What When Where Why*" model. Download the modified resource I've made below or read the Safety Action Step below for more.

Build Critical Thinking Skills: Safety Action Steps

1. Use critical thinking questions to practice evaluating situations together.
2. Teach your kids to ask questions about the media they encounter.
3. When they have questions that research will help answer, teach them how to vet the source and evaluate if it's reliable or has a vested interest in a certain perspective. For a detailed review of how to do that, explore resources from places like the News Literacy Project.[5]
4. Practice these steps as an adult in your online world and with friends, family, etc.

Most of us didn't get enough training during our school years.

Storytime: Recycled Disinformation

In 2020 when Joy was fourteen, something disturbing started to happen. We were driving in the car one day, and she asked about something she'd recently heard that I found preposterous. I don't remember the content, but what I remember most was being shocked. My head practically swiveled around on my neck, and I asked where in the world she'd heard this information – TikTok.

I explained what a conspiracy theory was, and I didn't think too much about it. Until it happened again.

Once more I was driving (our deepest conversations seem to happen in the car), and I asked her to do a quick internet search for me while my hands were on the wheel. I guided her through finding sources of information that provided proof, debunking what she'd learned.

The third time it happened, I was able to spot the pattern. The stories were recycled disinformation that had originally been passed around the internet during the 2016 election cycle. It has been proven that there were active digital disinformation campaigns by foreign bad actors to impact the outcome of our presidential election (U.S. Senate Select Committee on Intelligence).[6] It was effective, too, due to our lack of critical thinking and media literacy.

Each story she told me was a throwback, word for word, of stories from the lead up to the 2016 election. Joy and her friends were too young to remember them from the first time around, so they were ripe to fall for the same tactics.

We did this again and again, me sharing proven facts, prompting her to research the topic (a.k.a. receipts), and over time, she realized that I was more educated and trustworthy than the people she was listening to on social media. Of course, it helped that she could see the proof for herself, because what do moms know?

These repeated lessons and conversations were a golden opportunity to put critical thinking and research into practice. She got more of these lessons in high school, but she was already prepared to do her due diligence in information gathering by the time her teachers were giving her official assignments.

This was a huge win, and I felt like we'd dodged a bullet.

Critical Thinking in Real-Time

Research is a large part of my job. Sometimes it's formal, and I'm sitting at my desk poring over published academic papers; other times, the research shows up without any effort on my part. This data can appear when I'm scrolling social media or reading news from all over the world.

I learn about new scams, security risks, and concerning trends almost daily. I review the information, dig deeper into the concern, and incorporate this new data into my consulting services or presentations. I created a list of *Internet and Social Media Red Flags* to share with others in an attempt to prevent as much damage as I can.

Let me set the stage: you're hanging out, scrolling your feeds, and bam! A red flag. But it's likely you won't recognize that it's dangerous because you haven't been taught the signs. And you walk straight into the danger zone. In some cases, such as identity theft or interacting with a virus, you may not feel the effects right away. Some traps you've seen before, or even fallen victim to. I encourage you to talk about these traps with the people in your lives. When you share what you've experienced, what red flags fooled you, and what you've learned, you'll help others avoid the same fate. I'll go first.

Storytime: It Happens to Me Too

I fell for a dog story. There are many versions of this urban legend, with the details changing a little from version to version. Daisy was a seeing-eye dog with her human in the World Trade Center on September 11, 2001. She led her human down the stairs and out of the building to safety, then *heroically* went back to lead out more people. She saved over 900 people.

I'm a sucker for animals and heroes (who isn't?) and when I first saw this story sometime around 2012, I moved too quickly, shared the story to my Facebook feed, and didn't think twice. Not long after, I read a fact-checking article by Snopes[7] about Daisy. I felt like a dope.

What if critical thinking skills were activated? What if I was more observant?

I might have paused before circulating the article, because:

1. Service dogs never leave their humans unless they're experiencing a medical emergency – and that's only to get help.
2. The amount of time that it would've taken to move that many people out (while panicking) was beyond the time available.
3. The wording was designed to elicit the highest levels of emotion. Every version I've read after I learned the truth reads like a bad soap opera.
4. The hero theme is effective at luring us in, and that's why it's a common manipulation tactic. The word hero is used liberally in

every version of this "Daisy story" I've found.

I'm glad I fell for it because that means I can tell this story again and again to you and my audiences. It shows how none of us are immune to manipulation. And it reminds me, and hopefully you, how to spot red flags within the story.

The worst part?

I'm still seeing people share that same urban legend thirteen years later. I share the Snopes link every time I see it, and I let the person know I believed it the first time I saw it, too. I don't want the person to feel dense, because we do not live in a culture that teaches critical thinking skills or media literacy. I hope that my redirection can help them understand that the information is false and provide them with a reliable source to vet rumors in the future.

Internet & Social Media Red Flags

Red Flag: Accuracy Is Questionable
- Go directly to the source to see if you can learn more. When an organization is in an active crisis, they aren't likely to post information about it on their website, but they might on their socials. Move to the next step(s).
- Google the main words, adjust date filters to a recent timeframe to find current information.
- Use fact-checking sites to quickly learn if the information is true. I recommend: Snopes https://www.snopes.com[8] and

FactCheck https://www.factcheck.org[9] as a start.
- Coach your kids on how to listen to their inner voice or intuition. This is a survival mechanism that we are designed to use. Practice it.

Red Flag: Quizzes
- Don't take online quizzes! They are collecting and selling your data. Examples:
  o What type of Disney character are you?
  o What type of food are you?
  o What does your zodiac sign say about your personality type?

Red Flag: Heightened Negative Emotion
- Be very wary of posts that create a sense of panic, fear, anger, or intense worry.
  o The bad actor is banking on you panicking and moving too quickly.

This overrides your logic and critical thinking skills.
- You're more likely to click on the potentially malicious link or behave rashly by sending the post or information to everyone you're connected to out of a desire to protect others.
- Slow everything down and check in with your emotions.
- Do research on the topic before sharing with everyone.

Red Flag: Mentions of Sex or Nudity
- Avoid posts that include links to view a celebrity's compromising or leaked sexual content.
  - This link is malicious. It will steal your data or, much worse.
  - Train your kids to never click on links or stories like this.

- Train your kids to find you *immediately* if they receive a message saying the content is about them. They may have actually shared a nude photo and they'll be terrified to tell you. You need to let them know, no matter what, that you'll be there for them. *Do not blow your top.*
- Keep reading, we'll talk about how to handle this later in the chapter.

Red Flag: Hero Worship

- Keep moving when you see any post that uses hero worship or overt displays of love and support, and asks you to support by engaging with the content.
- What's the harm? You really do love and support them. There are a few reasons why not to interact or share:
    - Bad actors may harvest data from the comment section or from those

who have engaged with the post. If you interact, you're likely to be put in a "Sucker File" for future targeting.
- They include malicious links.
- They are planning on selling the account to another bad actor, and they're artificially inflating the engagement, shares, and interactions on the account or page to increase its value.
- Types of posts that fit into this category:
  - Like if you love your daughter. (We know you love your child. A like isn't necessary.)
  - Share to support a firefighter. (Donate at a pancake breakfast or other fundraiser and support them for real.)

Red Flag: Inspiration Porn
- Avoid posts that seemingly celebrate those living with disabilities.
- They share the same risks as Hero Worship Red Flag posts.
- "Inspiration porn" is a term coined by disability activist Stella Young (Forbes).[10] These exploitative posts may contain:
    - Sentimentality and/or pity
    - An uplifting moral message, primarily aimed at able-bodied readers
    - Objectification
    - Stereotypes or inaccurate information
- What's the harm?
    - I am an able-bodied person so this is not my story. But from listening and learning, I've come to understand the multiple opinions in the disabled community about the damage inspiration porn can cause. To

summarize briefly, objectifying people in both positive or negative contexts removes their humanity and creates barriers to empathy or compassion for their real problems and existence. Secondly, people with disabilities are simply trying to live their lives in an ableist society. They aren't here to inspire you to be grateful for what you have.

Red Flag: Terminal Illness, Support or Overcoming Obstacles
- Avoid posts that celebrate a person who is fighting a terminal illness or those who have recently overcome one.
  - If you know the person with the illness in real life and they have given permission, by all means, support them.

- o Posts encouraging taking action (give blood today!) are the best use of social media awareness in this category, but so rarely are they framed in this way.
- When donating to a fundraiser, confirm it's actually connected to the recipient to avoid fraud and theft.
- *Do not* interact or reshare one of these posts if you don't actually know the person.
- What's the harm? Reasons why not to interact or share:
  - o The photo may have been used without the individual's or family's permission.
  - o The person may have died, and the trauma of unexpectedly seeing a loved one's image used for someone else's advantage is a special kind of horrible.

- The story may not be true, and those who interact are helping to raise the visibility of this violation.
- Bad actors may harvest data from the comment section or from those who have engaged with the post. If you interact you are likely to be put in a "Sucker File" for future targeting.
- They can include malicious links.
- They may plan on selling the account to another bad actor, and they're artificially inflating the engagement, shares, and interactions on the account or page to increase its value.

Storytime: Shares That Hurt Like a Knife

Talia* is one of my closest and dearest friends. Her son Sam* was diagnosed with cancer at age eleven. His family regularly posted about him to update loved ones about his ongoing treatments and health, so seeing his photo in my feeds was common.

[Sidebar, Great Use of Social Media: They maintained a private Facebook Group so that they didn't have to repeat themselves over and over when asked about Sam's health. She invited people she knew to join the group, and they could stay abreast of important updates. We were able

to support, stay updated, and show up, as members of her wider community.]

When Talia's son died at age twelve, it was heartbreaking. I was on Instagram a few days after he passed, and I saw his photo. I was shocked to see it wasn't my friend's account that had posted the update, and I immediately let her know. Talia knew the organization that posted about Sam, but they hadn't asked if it was okay to post. She wished they had.

This is a reminder that the content you share, even with the best of intentions, may hurt people in unimaginable ways.

*Note: Names in this story have been changed

Want a simple list of dos and don'ts to review with your child about what's okay online?

Read on or download to keep. When talking to your kids about this topic, include the reasons why it's unsafe. "Because I said so" isn't enough of a deterrent. Give them logic and they're more likely to follow rules. They may also share this knowledge with kids around them and extend your protection to others; at least that's what happened for us.

DON'T DO IT

- Give their full name, birthdate, phone number, email address, passwords, physical address, or school name to anyone online.
    - Why: This information can be used to track, stalk, or steal the identity of the child.

- Give detailed information on the family and their activities.
    - Why: This can be used to threaten jobs, reputation, and physical safety.

- Send anyone you don't know pictures, explicit or not.
    - Why: Images can be used to impersonate, as source material to create AI deepfakes, or altered images portraying the victim as nude. Predators (or other kids with malicious intent) may use these deepfakes to threaten a child into giving further explicit content or engage in additional dangerous behavior when threatened with extortion (also called sextortion) or exposure. Children die by suicide due to these types of threats. Read more about *Sexual Exploitation* a bit

later in this chapter.

- Show off any parts of their body on camera, even if they know the person well.
    - Why: Images can be edited to appear fully nude. Same risks as above.

- Video chat with a person they just met.
    - Why: Predators commonly pressure victims to video chat within the first couple of days after meeting online.

- Click on links in emails from strangers.
    - Why: This is the most reliable way to get viruses, malware, or spyware onto a device or computer. Adults are the most at risk of being victims of this due to our high use of email. Younger children don't use email (yet), but teenagers are beginning

to. Teach them about it and keep telling them (McAfee).[11]

- Click on lonely links.
    - Why: Emails or texts arriving from known people that only contain a link are often due to the account being hacked or spoofed. The subject lines are often designed to worry you into clicking "I just saw you on this website/doing something crazy/this is so scary!" The bad actor will send this same email to the entire contact list of the account.
    - Recommended action: Ask the person in another way if they sent it to you before clicking on the link. If the sender wasn't hacked (this is rarely the case), let them know it's a tactic of hackers to send just links to discourage them from that behavior.

- Click on links in texts from strangers.
    - Why: Text or SMS phishing (also known as "smishing") is when a deceptively written text or SMS from an unrecognized number has a malicious link included. Smishing fraud is on the rise. You can use the "delete and report" function on your device or go a step further and report to your wireless carrier or within a call blocking app (FTC).[12]

- Respond to hurtful or disturbing messages.
    - Why: There's nothing good that will come from interacting with this kind of message. It gives the sender reason to keep it up if they get a response. Their goal is to scare, upset, or otherwise trigger the recipient.
    - Recommended action: See *Chapter 5 | Family Safety Plan* for more

guidance on what to do when this happens.

- Publicly share physical location or "check in" on an app.
  - Why: People of any age can be tracked, stalked, and abducted based on location information. This can happen quickly.
  - Recommended action: Parents, *please* stop sharing photos of your child's first day of school in front of the school sign or tagging the location. You are putting your child at risk! If you must share their first day in a public way (which I don't recommend), remove any identifying information about the location or school.

- Meet online-only "friends" in person.
    - Why: There is no reason a good person would want to meet your child after finding them online. This is the *most difficult* lesson to sink in. I don't have a magic solution, but I'd drill into them that, "NO GOOD OR SAFE PERSON WILL WANT TO MEET YOU IN PERSON AFTER YOU'VE MET THEM ONLINE. THEY WANT TO HURT YOU."

- Download apps without permission or from an unapproved app store.
    - Why: Kids with free rein of an approved app store may download all kinds of inappropriate apps.
        - Apple phones are a bit safer for this. It is possible to download apps from outside the official app store, but it's difficult.

- Android phones are less safe on this one. Google does not have a stringent app review process, and many of the apps in its official store are malicious or dangerous.
  o Recommended action: Lock your child's ability to download apps without you. See *Chapter 6 | Setup Parental Controls.*

DO

- Treat others as you want to be treated.
- Be a model of good behavior.
- Understand that there is *no such thing as privacy.*
- Use your cameras wisely, and block when not in use.
- Listen to your gut and your instincts.

- Know all of your connections. Remind yourself that there's at least one person on your friend list who is lying about who they are.
- Break off communications or friendship with a person who violates your boundaries or makes you uncomfortable. You *do not* have to stay friends.
- Remind your kids to talk to a safe adult when in doubt.

Storytime: Overexposing Children

Ah Facebook, the town square for adults. If you're like me, you've got lots of Facebook friends who are people you went to school with long ago. This is a great way to catch up with their lives today and stay in touch.

But even staying in touch can get weird, especially online. One of my Facebook friends was a woman named Julie* and we were in chorus together during junior high and high school. We weren't particularly close when we were young, but we'd found each other decades later, online, as you do. Over the years, I'd seen many posts about her family, her husband (whom I also knew from school), her life as a stay-at-home mom, and homeschooling activities of their six kids.

One day, I was at the mall, standing behind a group of people as they were paying for their parking. The mom turned around and recognized me. It was Julie. I noticed each of her kids, and as I looked from face to face, I almost called them by their names. I stopped myself because I realized I was a total stranger to them. I knew the names and many personal details of each one of these children. I felt creepy and dirty for knowing so much about them, even though we'd never met. I hadn't done anything wrong, but I was a witness to violations of their privacy.

Eventually, I unfriended Julie because I never got over that icky feeling. I wanted to set boundaries between myself and children who weren't able to consent to their overexposure. I couldn't help but think about her other online friends who had more than enough information about her innocent children to be dangerous to them.

Note: Names in this story have been changed.

Children cannot give consent.
Your child cannot legally
or emotionally give consent
to having their image
or likeness posted online.

Sexual Exploitation

When talking about internet use and its dangers, I must mention sexual exploitation. Though I'm not an expert on this topic, I encourage you to learn more about it because it's a growing problem. When young and inexperienced users are online, we must acknowledge and prepare for this very real danger.

According to the UN Refugee Agency,[13] sexual exploitation is "any actual or attempted abuse of a position of vulnerability, differential power, or trust, for sexual purposes, including, but not limited to, profiting monetarily, socially, or politically from the sexual exploitation of another. It includes but is not limited to exchanging money, employment, goods or services for sex. This includes transactional sex regardless of the legal status of sex work in the country. *It also includes any situation where sex is coerced or demanded by withholding or threatening to withhold goods or services or by blackmailing.*" This form of sexual exploitation that includes extortion is called sextortion.

Why am I bringing this up?

Many people (this isn't exclusive to being young) can find themselves in situations where they're at risk of being exploited. A common scenario our children might find themselves in is when they make a new "friend" on social media and their new friend suggests moving to direct messaging inside that site or to a whole new location like a messaging-specific app (private communications).

We've now entered the grooming phase. Grooming is when someone seeks to build an emotional connection with another person to gain their trust for sexual purposes or other forms of abuse. It happens both online and face-to-face. It can be a stranger or a person they already have a social relationship with. (Learn more about online grooming by Internet Matters.)[14]

Sexual exploitation, or sexploitation, on the internet can include: Sharing nude or partially nude photos or other media; discussing past sexual activities; live streaming and encouraging sexual behavior; sharing intimate details; and more. Any visual depiction of sexually explicit conduct or content involving a minor is child pornography, also known as child sexual abuse material (CSAM) (U.S. Department of Justice).[15]

Some facts about sexual exploitation:

- Almost one in five young people who shared nudes were either blackmailed, bullied or harassed to send more photos. Source: Internet Matters 2019 Report.[16]
- 30% of twelve- to fifteen-year-olds said they had been contacted by a stranger online who wanted to be their friend. Source: Ofcom 2020-21 Report.[17]
- In 2019 and 2020, Meta-owned platforms (including Facebook, Instagram, and WhatsApp) accounted for approximately 94% of all CyberTips sent in by the industry. Source: U.S. Department of Justice Technology Report.[18]
- In 2008, 15% of American youth reported experiencing sexual or physical abuse or high parental conflict in the preceding year. Labeled as "high-risk", they were disproportionately likely to be older teens,

Black, and/or not living with their biological parents. Source: Wells & Mitchell.[19]

How to Protect Your Child

We will see many of these tactics repeat throughout this book (in the *Inappropriate Content, What Kids Can Do* section of this chapter and in the next chapter on *Bullying* as well as *Chapter 6 / Monitor Activity*) but here's a succinct list when it comes to making sure your child is aware that grooming can happen.

Safety Actions Steps

Note: Adjust the following steps for age

- Reinforce that they can talk to you or another trusted adult should any concerning situation arise.
- Regularly talk to them about the digital spaces where they communicate with friends and ask about the kinds of content shared.
- Give examples of what healthy and unhealthy relationships look like, so they have concrete examples for comparison.
- Explain to them how easy it is to pretend you're someone else online, and that some

adults may wish to approach them and conceal their true identity.

- Remind them that even though they're "friends," the other person might not be who they say they are.
- Review their friends list, looking for adults or other red flags.
- Review their privacy settings to ensure they are set at the highest level and teach your child how to report inappropriate behavior.
- Know the security or lock codes for their devices so that you have the ability to review the content for yourself.
- Open every app, as some can be disguised to look like innocuous tools (such as calculators) that can be secret photo repositories.
- Read their chats or direct messages in every social app they use, including video games.
- Teach them how to say, "No" and help them practice it in a variety of ways. As a

society, we don't support people in saying no. We worry too much about making others happy and it's commonly accepted for people to disregard a no and instead, push for a yes. They need your help in learning how to say no. Their lives depend on it.

Watch Out

Keep your eyes peeled for these signs and investigate further should you see any of these behaviors:

- Being secretive about internet use and conversations.

- Increasingly spending more time on the web.
- Having secret devices or hiding how much they access sites or the internet in general.
- Using sexual vocabulary beyond what is appropriate for their age.
- Out-of-proportion emotional responses to being monitored, questioned, or restricted from internet use.

Report Abuse

If you discover that your child is being groomed or has been targeted by someone in-person or online, please review and engage in the activities below.

- If there is immediate danger, call 911 (U.S.) or the emergency number in your location.
- Reassure your child that it's not their fault.
- Report to your local authorities.
- Report to your state, territory, or tribe-specific authorities https://www.childwelfare.gov/resources/states-territories-tribes.[20]
- When there's possession, distribution, receipt or production of child pornography, report to CyberTipline (https://report.cybertip.org).[21]
- Use the Take It Down service to remove online nude, partially nude, or sexually explicit media taken of minors (https://takeitdown.ncmec.org).[22]
  - If you're an adult that was abused as a child in this way, I'm so sorry you experienced that. You can also submit a request to Take it Down.

- Track all your reporting activities with entity contacted, dates, times, and any follow-up actions.
- Find survivor support locally or view this list of national organizations (https://enoughabuse.org/get-help/survivor-support).[23]
- Listen to your child's fears, respond to their needs, and provide love and support. This can be an especially vulnerable process and time for both you and your kid.

Why report to so many entities?

If you don't want to go through that many steps, I understand. You're worried, stressed, and your child is impacted. Your child's safety is your priority, and you may want to simply address their safety and hope the rest of it goes away. It will require emotional energy to go through all of this. But ignoring something does not make it vanish. In fact, ignoring it might make the emotional impact worse. Taking action, exerting control, and demonstrating to your child that they are worth protecting at all costs will create a very important sense of security that they will carry with them. This is not the time to look the other way.

Your child is likely not the only victim and, if not caught, the perpetrator will continue to harm others. If the bad guy is a part of a large-scale operation, the number of victims could be high. By reporting to multiple law enforcement and support agencies, you increase the odds of that person being stopped before they hurt more people. Your reporting also adds to the data and information available as some of these organizations work together.

Plus, not all law enforcement agencies are the same. They work these cases differently, and they don't have an equal level of effectiveness. I've heard horror stories from parents over the years, and my heart aches for those families and victims.

My advice to you is now is the time to channel your ferocious protective energy and unleash it. It's *Mama Bear Time.*

Storytime: Love Hijacked

I was at a social media conference when I joined a small table of people sharing their professional expertise. I told the group about my background and afterwards a woman pulled me aside to tell me her child's painful story.

Her son attended high school in Minnesota, and he was in love with his girlfriend when she sent him a racy photo of herself via text. He was probably excited to receive it, and he showed it to his friends in a small group. One of the boys hijacked the private share by snapping a picture of the phone screen before sending it to the majority of their school.

Nothing moves faster than an explicit photo. That image passed around the phones and devices of other minors in this small and tight-knit community like lightning. The young woman was shamed, harassed, and ridiculed. She ended up moving to another school due to the damage to her reputation.

Your first instinct may be that she shouldn't have shared the photo. In today's world, a large majority of people using smartphones have shared an inappropriate photo or video, it's a simple fact. I have to redirect the focus and emphasize that she shared it with someone she loved and trusted. The person at fault is the young man that violated the trust of his friend and shared CSAM willfully.

Having "The Talk" about Porn

Every family handles this topic differently. As a parent, you need to know that sexual content is everywhere – in the media and on the internet. We can't shield it from our kids' eyes forever. At the push of a button, it's on the phones and devices of their friends, classmates, and others in their world. This means it's important to address the topic. The site "The Porn Conversation" may help you feel more prepared and a little less awkward: https://thepornconversation.org.[24]

Teens Sharing Photos

Inappropriate images of minors get shared all the time, sometimes by teens themselves. Like wildfire, a picture or video of that nature can spread all over campus or a community within minutes. This means that all of those people now have CSAM on their devices. Sometimes the victim doesn't know what's going on because the other kids are afraid to say something.

Sharing or possessing CSAM is a *felony*.

If your child ever receives content like this or is aware of it happening, they need to go to an adult *immediately* because law enforcement needs to be contacted to protect the victim. Once your child has informed you or another adult, remove the illegal content from the device.

We can't expect our kids to know what to do during an emergency, and that's exactly what this is. We have fire drills at schools and listen to instructions about where the aircraft exits are located every single time we fly. We need to inform our kids of the dangers around them so that we can teach them how to react, escape, and protect themselves.

Revenge Porn

The *Love Hijacked Story* could've followed a different path, and it too often does. Revenge porn is the distribution of sexually explicit images or videos of individuals without their consent and it falls under the larger umbrella of nonconsensual pornography. Often, a former partner exacts revenge by distributing what were private images to a wide swath of people.

Currently forty-eight states plus the District of Columbia and Guam classify revenge porn and nonconsensual pornography as a crime, usually a misdemeanor (Criminal Defense Lawyer by Nolo).[25] The consequences vary by state and can include jail time and fines. Make sure your children understand the risks, dangers, and potential consequences of taking explicit images and sharing them without consent.

You can't keep your child safe every minute of the day, and that's not the point. Our job is to keep them safe when they're small and teach them how to protect themselves as they grow. They will leave your nest; and when they do, they need to be prepared for the risks and dangers they will face.

CHAPTER 5

Bullying

When I'm invited to speak on a school campus, it's for one of two reasons. One, something terrible and tech-related happened. Two, the adults are concerned about bullying. I have a few thoughts (and some data) to share before I get into what you need to know to keep your kids safe.

Cyberbullying, or the use of electronic communication to bully a person, is often shortened to just bullying. When I'm asked to comment on the subject, usually on stage in front of an audience, I find it's a bit of a trap, often used to complain about technology at large.

"Technology is bad and it's bad for us. Just look at all the terrible things people say to each other online," is the standard angry statement. While I can't disagree with the fact that tech has lots of negative repercussions on our lives and health, people who make a statement like that aren't looking for a constructive conversation on the topic. I imagine that they want to destroy a computer reminiscent of a scene in *Office Space* (1999) where the main characters take their frustrations out on a regularly broken printer and beat it with baseball bats like a piñata. I find that bullying is the broken printer we like to smash, the gateway term leading to the demonization of technology.

> *Is bullying terrible?* Yes.
> *Is it something that shouldn't exist?* Yes.

Is the internet responsible for bullying?

No.

Bullying (without the cyber) is an active and everyday part of human interactions. We accept it in our families, our churches, our media, and our friend groups. Almost everywhere that people are talking about or to others, bullying happens.

You can hear radio personalities bullying celebrities for the way they look. Magazines at the grocery store have covers displaying, "She's so thin! We're worried about her." Then the next week, the same person is in the "Hot Beach Bods" edition. (I wish I were kidding about that.) Women in the public eye, as well as those who don't fit into a narrative we're comfortable with, have always been bullied and treated as less than human by "news" organizations of all kinds. This behavior isn't new. It didn't just come into being with the invention of the World Wide Web, smartphones, and social media.

I'm not excusing this abhorrent behavior. And yes, online, it's easier to bully people we've never met – and at a grander scale than ever before. My point is that the internet is a mirror of how humans act. What happens online is a continuation of the behaviors we see offline. Bullying is a wider societal issue that we can't blame the internet for.

When we verbally beat up people during mean-spirited gossip sessions over coffee or cocktails, our words disappear like vapor into the air. This allows us to deny (to ourselves and others) that the words ever existed, and any proof or opportunity to reflect on our bad behavior disappears with it.

But the internet is listening and recording. Online, our words are written in undeniable text, recorded on video, indexed by search engines, and searchable for years to come. When we are appalled that a person could treat others in such a way, we are also shocked to realize that our behavior has been witnessed, recorded, and that we may have to revisit it over and over. It's easier to look for someone or something else to blame.

We teach our children not to behave this way, but that doesn't stop us from engaging in it. Cue the blame and demonization. What I see is people being people, the way we've always been, and no one holding us responsible for how we hurt each other, regardless of the delivery device.

When I'm asked to comment on bullying by a well-meaning and understandably exasperated adult, my response is "we need to start by asking why bullying is an accepted behavior in our offline society, before we can solve the problem online." They are usually rendered speechless by my response.

We all want a button, an app, or legislation to fix the problem. Each one of us has a hand in this issue because we see or partake in bullying every single day. I'll admit that I've been a bully. None of us can escape responsibility, none of us is without fault. It's a big problem online but that's because we allow it to exist offline. Kids learn from the behavior they see around them, the good and the bad.

Technology has a part in this, but the biggest culprit is the fact that we allow it to happen without calling it out as our children grow up. Every time we witness bullying, whether it involves our family members, friends, or others, we need to label it for what it is and demand that it ends right then and there. Our kids need to see us stand up to it, call it out, and protect those who are being targeted. Even the people you don't like or agree with. No one deserves to be the recipient of cruelty or dehumanization. Young people need to see us stand up to it when they're little and at every age afterwards. This is how we stop bullying, cyber, or in-person. If you want it to end, we're all responsible for stopping it at the source.

Bullying Facts

Kids don't use the word "bullying" as much as adults. When they're little and adults are the source of most of what they learn, they'll use the word. But as they grow and listen to their friends and peers more, they prefer action-based words such as swatting, doxing, catfishing, being mean, pranking, punking, excluding people, intimidation, gossiping, spreading rumors, sharing private information, teasing, or causing drama. When you're talking to your kids about bullying, using a variety of words will improve the outcome.

There are many different ways that this can present itself: physical, verbal, cyber, and social (as in personal relationships rather than social media).

Here are some facts to know about bullying:

- Boys are more likely to engage in physical forms of aggression and bullying (National Library of Medicine).[1]
- Girls are more likely to be victims of both in-school and online bullying (CDC data via Pew Research Center).[2]
- Black teens are more likely to say they've been bullied because of their race or ethnicity (Pew Research Center).[3]
- Black girls are far more likely than other teens to say online harassment and bullying are a major problem (Pew Research Center).[4]
- Transgender and gender-questioning youth experience bullying at nearly double the rate of their cisgender peers (CDC 2023).[5]

But it's worse than ever, right?

Not really.

By no means do I want to downplay the real impacts that bullying has on our children, particularly those who are LGBTQIA+ or people of color, but our perception of an exponential increase in cyberbullying is partially a clickbait scenario.

Clickbait is a sensationalized news headline that encourages or baits you to click a link to an article, image, or video. Instead of presenting objective facts, clickbait headlines appeal to your curiosity or heightened emotions, of which anger and fear are the most effective. (Remember those red flags listed in *Chapter 4?*) Websites and news organizations that feature the upsetting headline are receiving financial gain (via advertising or other profit-sharing agreements) per click and viewer time spent on that article. It's in *their* best interest for you to open the article, and they will use ever-increasing levels of panic to keep you clicking.

Monetizing your attention isn't a new concept that originated with the internet or social media. You've likely heard the phrase, "If it bleeds, it leads" which may date back to the 1890s. Publishing magnate William Randolph Hearst engaged in what was called "yellow journalism" or sensationalism after noticing that stories involving horrific incidents gained more attention from readers than more tame stories. Think about any news publication and what events get the most coverage.

Why do we click on those headlines and articles?

It's a survival instinct. To survive, humans have had to learn as much as possible about things that are a danger to our lives. We are trying to minimize our own risk. In the media, it translates to reliable audience behavior. They put up something gory, scary, or that screams "danger!" and we'll look. Every single time. Over time, headlines have relied more and more on salacious wording to capture our attention.

Is the amount of bullying bad?

Yes, yes, it is, but your perception of how bad it is has been impacted by increasingly emotional headlines that are designed to worry you and, of course, get you to click.

Let's take a look at the graph below.[6] You'll notice that reporting of cyberbullying at schools has gone up over time. But look at the solid line for regular bullying. It's dropped 8% while cyberbullying has risen 8% over the 10-year comparison. I was surprised to see such a slow incline for digital behavior; you'd think it was a house on fire the way the media reports it. I looked again and combined both types. In 2009, we had a combined

31% of public schools experiencing bullying. In 2019, it's the same amount.

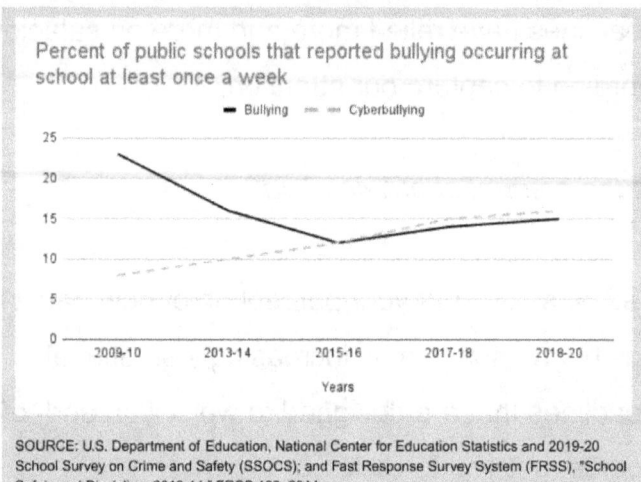

Did you notice this data is several years old?

There isn't more recent comparison data from reputable sources for the U.S. I did not consider any data from cyberbullying organizations because they have a vested interest in showing statistics that support their existence (this is critical thinking in action). Well-respected anti-bullying groups will cite scientific national or international studies, and the more questionable groups tend to create proprietary data from their own surveys. As long as I've been guiding families on surviving bullying, it's been normal to have data that is several years old.

Schools were completely upended during the national lockdowns in 2020. Kids were out of physical school for about two years, and the national studies and statistics have a big gap after 2020. That means the meaningful data will start back up in 2022 or 2023, and those reports will take approximately a year to be released.

When there's an incident, schools call me, and then I'm back on campuses, witnessing the panic amongst the adults. Understandably, they're scared for their kids and feel out of control. My job is to educate and guide them through the situation. I find the comparative data useful for diffusing the contagious anxiety that radiates off of parents and administration. I've been using yearly comparative data in every presentation I've given on the topic. I update as new data is released and it's the same story every time: no massive jump, at least not the way you'd expect from the way the news headlines are written. People struggle to reconcile the data with the hype.

Learn more about the current state of bullying by reading the *Teens and Cyberbullying*, a 2022 report by Pew Research Center.[7] Or *9 Facts About Bullying (2023)* also by PRC.[8] Pew is a highly reputable source that puts out data-heavy reports on a wide range of topics that the average person can easily read. StopBullying.gov,[9] provided by the U.S. Department of Health and Human Services, offers more resources and statistics.

Bullying on campuses is not on the rise, but it is still a problem.

Take a deep breath. Unclench your jaw. We'll get through this.

Signs Your Child is Being Bullied

It's our job to monitor behavior, health, and anything else that could negatively affect our kids. Your children may or may not come out and tell you that they're experiencing bullying or violence in their lives. Remember this activity can happen anywhere – at school, home, church, clubs, sports, online, via text – and it can cross from in-person to digital and back again. That's why the term cyberbullying isn't always the right way to describe this behavior.

The National Center for Missing & Exploited Children (NCMEC) has a great list of red flags that could mean your child is experiencing some kind of bullying:

- They stop using their computer or cell phone;
- They act nervous when receiving an email, direct message, or text;
- They seem uneasy about going to school (or other regular outing) and have increasing reasons not to go;
- They withdraw from friends and family; and
- They complain of physical ailments such as headaches or stomach aches.

Rather than ask your child if they're being bullied, ask them what's going on in their lives or at school.

Sample questions:

- What's going on at school?
- Are you having any problems with other kids?
- Is someone bothering you?
- How are things going with your friends?

When they start to talk, ask them to tell you more. Be supportive and gentle. If you can tell something isn't quite right, but they aren't ready to talk, tell them you're available whenever they're ready. You can suggest they talk to another trusted adult if they would prefer. Help them identify with whom they feel safest. If it's not you, don't take this personally. It will further demonstrate that you have their best interest at heart and engender trust.

Important Resources:

- Stop Bullying: https://www.stopbullying.gov/resources/get-help-now[10]
- Stomp Out Bullying: https://www.stompoutbullying.org/helpchat (ages 13-24 only)[11]

The previously mentioned National Center for Missing & Exploited Children (NCMEC) is an important resource to know. Visit their website, MissingKids.org,[12] for more resources. This is the most reputable organization when it comes to helping find missing children, reducing child sexual exploitation, and preventing child victimization. They work with every major law enforcement agency and social media platform. When inappropriate images of children are found on social media or the internet, NCMEC is a required organization to which tech platforms must report.

NCMEC's site offers a CyberTipline, resources on how to remove your own explicit content, AMBER Alerts, search for missing children, victim and family support, educational resources, recent data, and so much more.

Even if you do not need their information, you may find yourself in a situation where you can share it with others who do.

Storytime:
Sharing Resources with Community

I love sharing resources with people in the moment a need arises. It's something I passed down to my daughter Joy.

Several years ago, I had a social media contract with the local rape crisis counseling center. They posted a list of their resources and services, and I sent it to Joy, who then shared it via a group text with many of her friends. Who knows how many of them passed it along. With the multiplier effect, that one post may have reached hundreds of sixteen-year-old girls. I felt a strong sense of pride that day.

When I finished this book, I asked for Joy's consent to share about her childhood. She's now a nanny and babysitter, and she told me some fascinating stories about how she's using her knowledge of technology to help parents protect their kids. Not only did my parenting keep Joy safer, but it also equipped Joy with the knowledge to pass key information and resources forward in our community. Proud doesn't even begin to describe how I feel.

Hear directly from her.

Protect Your Child: Build a Family Safety Plan

I want to present you with two versions of a Family Safety Plan – one for the adults and one for the kids. When I work with schools, I often present to students during the school day and parents in the evening. I design the talks as companion pieces, giving all parties the same language to communicate with each other more effectively.

You and your child might work through a bullying situation together, but that won't always be the case. Remember, our job is to teach them how to be safe because we won't always be with them. They might help another young person through a difficult situation, and their action list below will stand on its own.

It's also possible that you are not on their "safe adult" list. I know this idea is upsetting, but remember when you were their age? Did you ever fight with your parents or not get along? There were several years of my teenage experience when my parents were not the people I would go to if I were in trouble. Let's make sure our kids have the tools they need, no matter who they go to for further support in these moments.

Think of the kid safety list like an emergency drill: children need to know the plan, and they need to practice it. Review the plan with your kids to make sure they know what to do when an emergency happens.

The Family Safety Plans below are adapted from advice listed on the U.S. federal government website StopBullying.gov.[13]

Family Safety Plan for Parents:

1. **Get Prepared Before Bad Things Happen**
    - Help your child identify at least two trusted adults in their life. These could be a parent, a relative, a school counselor, a pastor, etc.
    - Help your child identify their closest friends or their "crew." I expand on this in the next section.

2. **Notice Behavior**
   - Monitor for changes in mood or behavior and explore what the cause(s) might be.
   - Try to determine if these changes happen around a child's use of their digital devices.

3. **Talk to Them**
   - Ask questions to learn what is happening, how it started, and who is involved.

4. **Create a Record**
   - Create and maintain a record of digital or physical threats and actions with dates of events and useful details.
   - Take screenshots of harmful posts, content, and communications. Save links to any online content, if possible.

5. **Report to Authorities**
    o Report offensive content to the app or social media platforms to start the process of removal.
    o If the behavior happens at school, report to multiple school officials in this order: teacher, school counselor, principal, and superintendent. Keep records of your communications and let each person know who else you are reporting to; this helps keep everyone accountable.
    o Inform your state Department of Education. Schools regularly fail to protect their students in cases of bullying, and the more people you inform, the more likely you are to get assistance and action. If school officials aren't helping, see the * at the end of this list.

- If the behavior happens in a non-school environment, report to the responsible adult or decision-making body.
- If a child has received physical threats or if a potential crime or illegal behavior has occurred, report it to your local police.

6. **Block the Bully**
   - Wait several days rather than blocking right away.
   - Increase protection and privacy settings.
   - Review online friends with your child. Delete any online friends or followers that may be fake. These profiles are often where bullies are hiding.

7. **Support**
   - Sometimes public intervention is appropriate. Adults and peers can positively influence a situation where

negative content is posted about a child. This can include posting positive comments about the person targeted to shift the conversation in a positive direction. It can also help to reach out to the aggressor as well, to express your concern about their well-being.
- Provide your love and support to the child experiencing bullying. Make sure they understand that they aren't at fault. They might need professional mental health services to counteract the damage inflicted.

*There are no current federal laws that apply to bullying, but you can learn more about the U.S. Department of Education's Office for Civil Rights (via the U.S. Department of Justice's Civil Rights Division) and understand the options available to you and your child on StopBullying.gov.[14]

Family Safety Plan for Kids:

1. **Tell a Safe Adult**
   - Identify at least two trusted adults in your life, such as a parent, relative, school counselor, pastor, etc.

2. **Don't Respond**
   - Engaging with the bully is adding kindling to the fire. We want to starve the fire and give the bully nothing to work with.
   - Delete the apps from your phone or device, or block the offending website, rather than deleting the accounts. The bully can see you've deleted the account (kindling for the fire), but they can't see that the app has been removed or the site blocked (lack of activity or response starves the flame).

3. **Create a Record**
    o Create and maintain a record of digital or physical threats and actions with dates of events and useful details.
    o Take screenshots of harmful posts, content, and communications. Save links to any online content, if possible.

4. **Report to Authorities**
    o Report offensive content to the app or social media platforms to start the process of removal.
    o If the behavior happens at school, report to multiple school officials in this order: teacher, school counselor, and principal.
    o Keep records of your reports.
    o If the behavior happens in a non-school environment, tell the adult in charge. If they're not a good option, go to your safe adults for help.

- If you've received physical threats or if a potential crime or illegal behavior has occurred, report it to your local police.

5. **Block the Bully**
   - Wait several days rather than blocking right away.
   - Increase protection and privacy settings.
   - Delete any online friends or followers who might be fake. These profiles may be where bullies are hiding.

6. **Involve Your Crew**
   - Send up your "I Need Help" signal to your friends.
   - Talk to them about what's going on and how you feel.
   - Get mental health support if you feel it would be helpful.

Building Community and a Culture of Empathy

People are more powerful collectively than alone. We need community for health and survival because, just like predators on the savannah, bullies are more likely to target those who are isolated. Encourage your child and their friends to be a team, a crew, that protects each other. We can change bullying behavior, starting with our kids and their friends.

Here are steps to help kids protect and care about each other:

- Communicate with the parents of your child's friends about your goals in setting up a team meeting.
- Invite the friends over and serve snacks they'll like (it helps break down their walls).

- Seating matters. Aim for a side-by-side conversation, such as on a couch, rather than face-to-face, like around a dinner table. Kids are more prone to talk if you aren't staring them in the face. And the room doesn't have to be silent; it's better if it's got some regular noise. You could be watching a movie, doing an art project, or playing a video game. This feels less like "you're in trouble" and more like a safe space. This is hard, make it easier any way you can.
- Encourage a group conversation about what behavior is cool and not cool. Expect some disturbing stuff to come up during this talk. Stay calm and ask questions, but be mindful not to interrogate.
- Help them identify a "bat signal," a way they can tell each other that they need help or that one of their crew has been targeted. Giving the signal a cool name helps it feel more like a secret club.

- Encourage them to stand together as a team to surround each of their friends and protect them in digital and physical environments. Explain how words can be a barrier, and that they can also create a physical barrier between the targeted person and the offender, without violence.
- Help them identify adults whom they can find to begin the process of a larger intervention.
- Let them know that their parents, relatives, school officials, and authorities care about their well-being and have their best interests at heart. Tell them that it's possible the first adult they speak to may dismiss them (it happens), and to persevere to find an adult who will listen and help.
- Remind them that if they witness someone outside of their circle being targeted or bullied, they can help that child too, by

talking to them directly and speaking to an adult.
- Practice (or role-play) what their "bat signal" is going to be, what they should do if one of their team is in trouble, how they can help, and how they can support each other emotionally.
- Ask about how it might feel to be targeted so they can identify the complex emotions involved. Perform a 360-degree role-play of the bully, the victim and the observer. Walk through all perspectives with them and how each might feel so they can stretch their emotional intelligence and understand the impact this negative behavior has on everyone involved.

The final part about emotions is important. We all need help feeling emotions and putting words to how we feel inside our bodies. The more you can invoke the emotional impact of bullying, the better your odds are that this young person will have empathy for the victim and decrease the odds of them being the perpetrator. We never want to believe that our kids are the bullies, but sometimes they will be. It all starts with us, the parents and caregivers, giving our kids support and being there for them, as best we can.

CHAPTER 6

How We Got Here

We made it through to the other side! I bet you're a little stressed and overwhelmed. Well, you did a hard thing by facing scary information, getting informed, and staying committed. I'm proud of you. Let's get to the good stuff.

You can't control how others behave, but you can control your response.

Where Your Power Lies

The above phrase may sound familiar. While helpful, it would be better if we acknowledged that previous trauma may override our logic and impact the way we respond to a situation. A more accurate phrasing would include that we can only *manage* our own emotions, feelings, and responses. It's not as catchy but it's inclusive and trauma-informed.

You don't control what happens on the internet and you never will.

I find this phrase freeing, but not everyone does. You're probably saying to yourself, "This lady named the chapter 'How to Regain Control,' so what gives?" You don't control the world; your power is limited in scope, in some ways. But expansive in others.

Here's how I summarize it:

You control internet access
You control the devices
You control the timeframe

Unsure of how to implement this power?

Fear is a major part of the internet as well as the modern-day human experience. Let me show you tactics to flex the power you have.

Would you like to know which button to push to protect your children?

If you've read the book straight through up until this point, you already know that's a bit of a trick question. I worded it just so to catch your attention. Just like good parenting in all respects, protecting them is a collection of small choices that you make every day.

You Decide When the Time is Right

Remember that I mentioned COPPA creates that hard line at thirteen years old for kids creating their social media accounts?

With COPPA, federal law states that social media platforms cannot allow children under thirteen to use their services, nor can they collect information from them after thirteen until they are eighteen, unless it's confidential and secure.

But websites are a whole other deal. Your child may be using the internet, even in class, long before the age of thirteen. Schools are quite good at fencing children in safe places when they're using the web for lessons, so I wouldn't focus your energy or concern there. Your kids cannot use social media on school-approved computers.

Those sites are blocked (even from Wi-Fi) on every K-12 campus I've visited.

Storytime: My Family's Timeline

You don't have to allow your children to use the internet just because they've met the minimum legal age. In my Twitter story time, my daughter was eleven when she started trying to use social media. Because I get asked this a lot, let me show you the timeline for how her digital footprint grew over the years. This was our timeline, yours will be different.

Age 0-6

- Streaming television on one TV in our house, for a limited time
- Occasionally used a parent's phone, not for prolonged periods

Age 6

- Got Mom's old-school iPod with a child-appropriate song library and no connection to the web
- Started using Mom's iPad with voice search
- Only a few apps were available and loaded on the tablet
- No web browser
- No ability to download anything
- YouTube was available, only with all the parental controls turned on (pre-YouTube Kids)
- Mom monitored YouTube (via search, activity history, and suggested videos)

Age 9

- Got an iPod Touch for her ninth birthday*
    - (Mom and Dad disagreed on the timing; Dad won this round.)
- Kids in her fifth-grade class started getting hand-me-down phones
- Peer pressure and bad behavior among all kids were the result

Bonus: having a device to take away became great leverage for behavioral change

Age 10

- Got a Chromebook for schoolwork

Age 12

- Begged for a phone
- Got Mom's old iPhone 6+
- All the parental controls were turned on
- Limited apps were loaded on the phone
- No ability to download anything

Age 13

- iPhone 6+ got damaged
- Threatened to get her a flip phone; it was sweet drama (See *Storytime: The Flip Phone* in *Chapter 6 / How to Regain Control*)

Age 14

- Got Dad's hand-me-down iPhone 8+
- All the parental controls were turned on
- No ability to download anything
- Downtime and app time limits were turned on

Age 14*

Context: COVID-19 lockdown. We entered a whole new world of tech usage.

- Negotiation for social media tools
    - Choices were Instagram, Snapchat, or TikTok
    - Snapchat was approved for two reasons: It was messaging and not

on the web, and I could monitor who her friends were
- TikTok was approved for a short time, then revoked due to increased concern around pre-election content. (See *Storytime: Recycled Disinformation* in *Chapter 4 / Staying Safe Online.*)
- All the parental controls were turned on
- No ability to download anything
- Downtime and app time limits were turned on

Age 17

- Removed parental controls after proven good behavior
- Could download apps without parental permission

When it's all written out, it looks easy. Was it? Not even close.

The timeline does not include all the fights with an upset preteen and later, teen. My spouse was on board with my hardline approach, deferring to my industry background. But it was just like any other thing we do in parenting: paying attention to our kids' needs, exercising patience, and staying disciplined to make good choices even when we were tired. And it was so tiring.

Did I just want to let my daughter have the devices without watching her like a hawk?

Yes. And sometimes I got lazy. But when I got back on my monitoring, I would be reminded of why I needed to stay on top of it.

Somewhere around sixteen, it clicked for Joy. She understood why I was making the choices I made, and she understood that I was protecting her – that it was my job. I did my best to explain *why* I was setting boundaries, *why* she'd get in trouble for breaking rules, and *what* the risks were. She saw that I enjoyed technology and social media, that I wasn't being a tough mom out of spite, or that I was a hater. I was being a good parent. She's grateful that I put in the effort and I showed my love, even when it was hard.

Throughout the years, people would ask if my child was using a certain piece of technology. They were always surprised to hear me say "No." People expected my kid to have the latest and greatest tech at the earliest age, but that was never the case.

When Joy was still little, I read a study that stated children didn't receive any benefits from using technology until second grade, as far as "getting ahead" when in school. I knew that Joy would have plenty of time for technology. Ask anyone who works for a big tech company; their kids are usually super unplugged and attending Montessori schools. We know what it does, and many of us do everything we can to preserve their childhood years, keeping them safe as long as we can.

Learn About Technology

Almost all adults lack basic education regarding the tools we use every single day. One reason for this is because much of the technology was developed after we left school – and we wouldn't have learned much about it there anyway. Kids today don't have it much better than we did, when it comes to a lack of education around online safety.

Where do adults learn the skills they need to educate their children and help them navigate the digital world safely?

This question is why I started teaching. The following is a list of *free resources* that will help you learn the basics of technology. It all changes quickly, so the process of learning never stops. You'll never "arrive" or know everything. That's not the point. We're trying to learn and understand enough to protect our children.

Here is a compilation of useful resources for your learning journey.

Learn from Google
- Search "How do I _____?" Fill in the blank as necessary and adjust your phrasing until you get what you need. Adjust the timeframe of the search results because social media and technology change quickly.
- Use any search engine you like, but if you're using Google, click on *Tools→Any Time→Past Year*. I use this to learn about new technology and recent changes.

Learn from YouTube

- There's a tutorial for anything you could want to know. In fact, while the #1 category is music, the #2 category on this platform is education.
- Just like when you adjust the date in Google Search (see previous paragraph) for a more current result, look for videos that are as recent as possible with a high view count.
- Fun fact: Google owns YouTube. Google is the #1 search engine while YouTube is #2. When you search for phrases in Google, you'll get YouTube videos in the results because they're the same company.

Learn from GCF Global Learning

- This site was created by Goodwill Community Foundation (GCF) to support those experiencing barriers to work. It's designed specifically for adults to learn

anything we need to know about technology and work. I love it!
- GCF has lessons on hardware, software, apps, and social media information. Available in English, Spanish, and Portuguese. https://edu.gcfglobal.org.[1]

Explore ConnectSafely
- Features guides for educators and parents. Guides are short and sweet, printable and online. https://www.connectsafely.org.[2]

Take a Lesson at an Apple Store
- Apple has an amazing array of in-store classes. I took one on Privacy and found it valuable. https://www.apple.com/today.[3]
- If there isn't one near you, read more in the *Your Mobile Phone Store* paragraph.

Understand Android Device Options
- While Apple devices are all made by Apple, Android (Google) devices are made by a

variety of manufacturers. That means there isn't one single physical location where you can get assistance for your exact device, but you can learn about your Android operating system (OS).

- To learn about your device, your best bet is to use a combination of the items listed here and/or visit your mobile phone carrier's physical locations.
https://www.android.com.[4]

Visit Your Mobile Phone Store (also known as a Phone Carrier)

- Check your service provider's website (AT&T, Verizon, TMobile, etc.) and see if they offer in-store or online education.
- This is where you'll find hands-on assistance for Android devices, or if you don't have an Apple store near you.

Look at Your Device Manufacturer Website

- For all other technology equipment such as game systems, smart appliances, and wearables, you'll be using a combination of the manufacturer's website, search, and YouTube.

- Look for official brand channels and those that are dedicated to education around any topic. There are lots of people who create whole businesses and content engines around education for a particular piece of technology. Look for those with lots of subscribers and high view counts.

Visit Your Local Library

- I'm a true library geek. I can't say enough good things about the library! They are full-blown information centers. Many of them have grant funding that's dedicated to upgrading and offering the newest technology for free to their patrons such as hardware, software, classes, and much more.

- Visit their website or talk to a librarian on the phone or in-person to learn about their options.

Explore Community Learning Options

- Most communities have recreation and parks departments in their city or county government. Check out the community college near you. Take a look at the websites for either and see what adult learning they offer. These classes are anywhere from a couple of hours to weeks or months long.
- The barrier to entry is low since it's just like buying a ticket. The commitment to learning for adult classes isn't the same as registering for a traditional college class. Adult learning options are low-cost, grade-free, and everyone is welcome.

- Note: Many of those budgets have been cut, or demand has dropped. These options are almost non-existent in my community today or I'd still be teaching tech classes at my local community college.

Explore LinkedIn Learning (Access for Free!)
- You will need two things: a library card and a free LinkedIn account. Check to see if your state or library system has an agreement with LinkedIn where they offer LinkedIn Learning for free to library users.
- As a Californian, my library is a part of this program, and I only needed to log in once to have ongoing access to this service.
- LinkedIn Learning normally costs $29.99/month at the time of this writing. Enter your library card number at the site I've listed here and see if your library system is in partnership with LinkedIn. https://www.linkedin.com/learning-login/go.[5]

Ask Your Kids to Show You

- Ask them to show you once. Ask them to show you again more slowly. Ask them to watch you while you try to do it yourself. Try one more time while they're with you. Keep practicing.
- The act of adults asking their children for help is a big deal for several reasons: (1) You're being vulnerable; (2) You are showing that you respect their knowledge; and (3) It can help build stronger bonds with your kid by asking for their help.

Increase Your Control: Safety Action Steps

By now, you know that there is no single fix to achieve the result you're looking for. Online safety requires a combination of effort and a lot of trial and error. You may not use every one of these steps, but it would be ideal if you did.

Set Boundaries & Expectations

I'm pretty sure your parents or caregivers didn't do this when you were young, but now is the time to set the new standard. Sit down and have a conversation with the whole family about what is and is not acceptable behavior between people when using technology.

If you don't know where to start or you are afraid to miss something important, take a look at this resource for what's sometimes called a "cell phone contract" https://connectsafely.org/contracts.[6] They have several kinds that cover use and expectations for children, teens, parents, and the whole family.

You can print and download the contract or simply use it as a guide as you build your own. The most important part is for you to map out your expectations of behavior, communicate that to your family members, and then set ground rules and consequences.

Don't think this is a one-time deal. You'll have an initial conversation, then follow-ups as things change or misunderstandings arise. When it comes to technology (or raising kids), we've got to reiterate and repeat.

Reality Check

You don't have to allow your kid to have a cell phone! It's true. They'll need access to a computer for school starting somewhere around eight, but they don't have to have their own devices. Having access to a machine that the family shares is a compromise. I was shocked to see that 24% of children aged five-seven in the U.K. have their own smartphone (Ofcom).[7]

It doesn't have to be this way, it is not inevitable. You have the final say.

Just as in any consenting situation, you can retract the consent you gave your child to use technology. You can call it a time-out, a reframe, a tech-free choice – it's your call.

Monitor Activity

Use a combination of monitoring tactics and remember: you're the adult in this situation. It's your device and you allow them to use it.* When you have your family conversation about expectations, you'll let them know that making devices available for review at any time is a condition of access privileges.

Foster parents, this is for you:
For years, I educated foster parents on technology use. There's a stipulation that a child in your care must have access to a phone. This doesn't mean they have to possess their own phone or have unrestricted access to one. I heard a few stories about foster kids twisting this one in their favor.

Check with your case manager, but access to a phone when they need it is different than your ward convincing you they must have a dedicated cell phone with no constraints. That's not the intent of the rule.

In my house, every single password and security code for devices that my child had access to was set by me, and she couldn't change them. This is a standard parental control setting and by doing a quick web search, you can learn how to do this yourself. If you need help, a tech at the Apple Store or your phone carrier can help you.

Write down all logins in your chosen password manager. Keep track of this information! It will allow you to monitor so much more easily.

I also recommend using Two Factor Authentication or 2FA. This is not the highest level of security available (passkeys are even better), but considering how much hacking is happening today, I say turn it on everywhere you can. When you log in to a website, app, or account, you'll be asked to enter a special login code that you receive via text, email, or an authentication app. I recommend using an authenticating app, such as Google Authenticator, in case the device is lost or stolen. Learn more about 2FA: Kerry Rego Consulting.[8]

Are you worried about invading their privacy?

There are a lot of layers here. Do what feels right. Think of what you wanted and needed at that age, not that you should have had it all, but at least pause and reflect.

As people, your kids want their privacy, and I know you want to respect that. But they're looking for privacy *from us* to a certain extent. As all young people do, they're looking to socialize, build relationships, and find their place in the world. They're trying to figure out how the public world works, and most places – jobs, bars, nightclubs, restaurants – are off-limits to them. (School and extra-curricular activities aren't always their choice, so they don't count.) The internet is one of the few places they have to themselves. They need to practice being socialized humans in every way they can.

Think about the "Keep Out" sign many of us posted to our bedroom doors when we were kids. Kids in any day and age want (and need) privacy, but you also need to know if they're building a bomb in their bedrooms. It's not out of bounds for you to monitor, but be mindful of going too far.

You can listen in on their conversations, but that doesn't mean you *should*. In our fear and desire to maintain control over every detail of our child's lives, please don't become their jailer. Find the balance between trust and guidance.

Monitoring Without Breathing Down Their Necks

How often should you check?

You'll find a frequency that works for you. I like a lot of monitoring at the beginning, then doing random spot checks. But to be honest, I got lazy. Over time, I would decrease how often I checked Joy's activity and question her less and less, until she went off the rails. As soon as she'd get caught doing something wild, I'd get tough again, increase the parental control features, monitor more regularly, etc., then my reviews would taper down after a while. We went through this cycle over and over for years. Eventually, she learned and it got less painful. But it's work, I'm not going to sugarcoat it. This is parenting. It's what we signed up for.

One tactic that helped me monitor regularly was lying down in bed or on the couch. I found that when I was comfortable, I would do a better job. I took my time, opened more apps, and was more thorough. Often, I'd start after she went to sleep, and if I found something I didn't like, I had to wait until morning to discuss it. This meant I could process my emotions, approach the conversation calmly and with more empathy, and follow a list of notes for our sit-down talk.

Use parental controls or apps to create guardrails and check behavior from a distance. I'll go deeper into this in a bit.

Ask to see their devices (all of them!) and scroll through their apps to review their communications, activities, and search behaviors. Ask questions about things you see, and ask for context in conversations that don't make sense. Learn about current trends and look up slang terms you don't recognize. Ask your kids and also look for yourself. Sometimes kids will lie about what something means so it's smart to double check.

Here are two resources to help:

- Chat & Text Abbreviations
  https://messente.com/blog/text-abbreviations[9]
- Social Media Glossary
  https://later.com/social-media-glossary[10]

Do you read every word?

I didn't. I wanted to provide Joy with a level of privacy, and I told her exactly that. I said I wasn't trying to invade her space, and that my job was to keep her safe. This meant I was looking for unsafe behaviors and red flags. I scanned her texts, DMs, comments, etc., and concerning words or emojis jumped out at me. That's when we had conversations.

There was one time I was scrolling through Joy's phone when she was twelve or so, and I noticed some red flag words in an iMessage game she was playing with a boy. I slowed down and backed up a little to get the full sense of what was going on. I took screenshots of troubling sections and sent them to myself (in case she deleted them) and wrote down my concerns. The next day, we sat down at the kitchen table and went through it one item at a time. It ended up that she was flexing; she was trying out a new way to express her personality, like trying on clothes. It was out of character for her, and because I quickly noticed what happened and we talked about expectations of behavior, she learned two things right then. One, I was watching her, and two, what she was doing wasn't okay.

Where to Monitor

Even if it's banned, in your home or by the government, kids will find a way. We can't leave it up to digital safeguards, we are the ultimate safeguard.

This is a suggested list, but you might think of more elements to investigate. If you're using a monitoring app of some kind, built-in or external, you can see where your child spends their time. Once you know that, you can prioritize what you review. Ask your kid for help to find where some of these are located, but consider using a reliable adult-created source for accuracy.

Places you may need to monitor:

- Text messages
- Chat rooms

- Social media comments, given and received
- Social direct messages (DM)
- Followers, friends, and subscribers
- YouTube viewing and comment history
- Search engine (Google, Bing, etc.) history
- Screen time and "My Activity"
- Games with chat functionality

Set Up Parental Controls

There are so many parental controls now! Everything is much easier to use than when I first began monitoring my kid's digital behavior. You can monitor devices, computers, internet access, time spent, social media use, gaming, and much more.

Many of these new options are due to pressure from local and regional governments. I can't keep up with the changes and neither will you. Just know that it's generally trending in the right direction.

There's a spectrum of regulations brewing: outright banning, age limit restrictions, notification suggestions to warn of dangers to the health of minors, and more. It's fluid and broad. But most of the legislation gets held up in court because they can't figure out how to enforce it without violating our Constitutional rights (here in the U.S.) or because it's technically too complicated to implement.

Even though governments and schools are gaining steam in the protection space and breathing down the necks of the tech companies to do what's right, the bottom line is that it's up to you.

Please, please use parental controls.* This is one of the most important actions you can take to protect your kids.

> "One of the things we do find ... is that even when we build these controls, parents don't use them." Nick Clegg of Meta (The Guardian).[11]

It's frustrating that so many parents don't use these controls, because they do a good job. And they only work if you turn them on. No one will protect your kid the way you will and no company should be expected to take on the role of parenting. Turn them on.

You will find all links in the Resources| | Tech Support & Parental Controls *section but some of* the tech support links are long and unwieldy so they only appear there.

Computer Hardware & Operating Systems (OS) Software

A computer system has two basic components: hardware and software. The hardware is the plastic housing of the machine that you can rap your knuckles on, as well as peripherals you can touch, like monitors, keyboards, and mice. The software or operating system (OS) is the coded set of instructions that enable the hardware to do specific tasks like print, create documents, edit media, access the internet, etc. The OS is what you regularly update as the engineers add improved features, functionality, and privacy settings while fixing flaws. Update your OS to keep your machines running smoothly and safely.

There are many names for personal computers, such as PCs, desktops, or laptops. These are considered traditional machines and don't include handheld devices such as mobile phones, tablets, wearables, or gaming systems.

The computers in your home, office, school, and those your children use will fall into one of these categories:

- A PC that runs on Microsoft WindowsOS
    - The hardware manufacturer of the computer is irrelevant. When I ask, people often say "I have a Dell or an IBM." Doesn't matter what company made it; it's a PC running Windows.
    - Review the parental controls.

- An Apple computer that runs on MacOS
    - The hardware is manufactured by Apple. Their computers are called Macs (short for Macintosh, a type of apple).
    - Review the parental controls.

- A Chromebook that runs on Google ChromeOS
    - The hardware manufacturer of the computer is irrelevant.
    - Review the parental controls.

All three of them have parental controls. I recommend you tour the options to get oriented. Get in there, poke around, and make changes suitable to your child(ren). Make it a regular thing (once or twice a year) where you go back and check to see what improvements and changes the manufacturers have made over time. Confirm that the settings continue to be your preference. Bookmark or otherwise save these parental control website links. Over the years, they will discontinue one page and start it up in a new place (this is so annoying). If that happens, simply do an internet search for "parental controls for ___ computer," fill in the computer or OS you're trying to protect, and you'll find more up-to-date information.

You may go up or down in your strictness depending on your child's needs, their mental health, or how they're doing in school. I like to line up these kinds of tasks with other major tasks such as changing my smoke detector batteries at daylight savings time. Consider setting appointments on your calendar whether monthly, quarterly, or January/July. If you don't decide on a schedule, you'll forget.

Is Apple or Android better?

Control settings are easier if you have Apple products or all have Google/Android and PC products. It's smoother to sync and control multiple devices from a central family share dashboard. Trying to cross brands is a big pain. Avoid it if you can.

Parents often start their kids with Android devices because they're less expensive than what Apple offers. Once they get the hang of not losing or breaking their tech, many parents tell me they switch their kids over to Apple, because that's what they have, and the whole family can be linked together. We're an Apple family, and it was easier than the alternative.

I recommend Apple products in every category (I know the PC & Android people will disagree) because of the control they have over the environment. Apple has a closed ecosystem: they manufacture all their products and monitor the software available in their App Store. Google has an open ecosystem: they allow different companies to develop hardware that runs their software and operating systems, and they have done very little vetting of software in the Play Store. The lower level of quality control means there is a higher instance of virus-laden apps and dangerous activity in the Google/Android product line.

Apple Devices

Includes Macs, iPhones, iPads, Watches, and more

- Use parental controls (iPhone & iPad).
- Check your cellular provider's website for additional apps and resources.
- Visit your local Apple Store or mobile phone store for human assistance.

Set screen time limits, set privacy and content restrictions (they have options similar to movie ratings), prevent all App Store purchases, prevent certain kinds of web content, restrict Siri and Game Center, and more. Next to monitoring their behavior manually, using these controls is easily the most important thing you can do to protect your child.

There isn't a central page for Apple Parental Controls with links to the different devices and needs. This is frustrating. I've given you the link for the most common needs for iPhones and iPads. To find additional instructions, search the web for parental controls and the device you'd like such as "parental controls for Apple Watch."

Joy told me to mention that when you set your child's age in the Family Sharing setup, they won't be able to override the age restrictions.* This means that when you set their age at the beginning, they won't be able to download content meant for older age ranges. It will prompt "Ask to Buy," and you will be prompted on your device to approve or deny the request.

*This is a perfect example of kids teaching parents how tech works! She's regularly showing me something I don't know. Be open to learning.

Google Devices
Includes Android phones, tablets, and Chromebooks
- Setup parental controls with Family Link.
- Use the Monitor Activity portal.

- Check your phone carrier or handset manufacturer's website for additional apps and resources.
- Visit your mobile phone store for human assistance.

You can block or allow certain apps or websites, monitor the apps they use (and for how long), and lock devices when it's time to relax, study, or sleep.

"Safe" Phones

There are phones designed just to keep kids safe. There are many options, but four brands consistently come up: Bark, Gabb, Pinwheel, and Troomi. I cannot recommend one over the others, but you should do your research. I found several sites that compare what these phones can do.

Get settled in, take some notes, and be an informed consumer the same way you'd research any other important purchase. I'd put all the info in a spreadsheet so I can compare many factors. Here are some suggestions as to what you'll be comparing: unit upfront price, monthly packages or ongoing costs, service coverage, security settings, and customer service.

What can't they do?

Be a parent to your child. Only you can do that.

Storytime: The Flip Phone

My favorite "parenting with tech" story is the one about the flip phone.

You've probably heard the term "dumb phone" which is a common term used to describe a phone that isn't a smartphone. People also use flip phone for these devices, but not all of them flip open (remember the old school Nokias?) The technical term is "feature phone": the manufacturers had to come up with a name that differentiated their products from their smart siblings without using a derogatory term like dumb.

Anyway, when Joy started begging for a phone, I told her she wasn't old enough – this worked for a long time. But I began to grow nervous about us not having a landline and the short periods of time when she was alone at home. I knew she needed some way to reach us if necessary.

Finally, when she was twelve, I gave her my old iPhone 6+. It was more locked down than Rikers Island. She couldn't do practically anything with it (she is my only child and I work in tech – I was paranoid). It had *all* the parental controls, very few apps were installed, and no ability to download anything.

One day, it was accidentally destroyed by water, and an opportunity presented itself. I had been threatening to downgrade her to a flip phone, and I realized I had the chance to stall forward tech progression.

I marched into a cell phone store and asked for a feature phone. The tech that greeted me at the door had no idea what I was talking about. I rolled my eyes and taught him something about his job by explaining it to him. The next tech I talked to asked what I needed.

I said, "I want the most prehistoric phone you have."

I explained my devious plan to him. He laughed and unearthed a dusty box from the back room, literally the only one they had. We snickered as he connected it to the network, and I had one of the most satisfying days as a parent that I can remember (YouTube).[12]

You already know that my kid *hated* it. But it bought me time. Yes, it was embarrassing, and yes, she was mad at me for foisting this tech upon her. She wanted a phone like her friends had. But she damaged her last one and had broken some rules, so this was the compromise.

Joy learned that with phones comes responsibility. She had to carry it, answer when I called, and respond to my texts. She did not perform well, and that was actually the point. After months of us going around and around about her showing responsibility to earn a new smartphone, she learned and became more accountable.

The end of this experiment was when our cell phone carrier removed her ancient device from our cell phone plan for "incompatibility," and it no longer functioned. I'd bought maybe six more months of low-tech time and taught my kid a valuable lesson.

FYI, you can also get your kid a feature or flip phone as their training wheels. Heck, you don't have to upgrade them to a smartphone if you don't want to. This is an option.

Joy does not approve.

This story is never not funny to me. Call me petty.

(Watch a mad Joy talk about being forever scarred by the Flip Phone Incident.)

Cell Service Provider

Many companies provide mobile phone service. Some examples include: AT&T, Verizon, Mint Mobile, TMobile, Consumer Cellular, Sprint, and Boost Mobile.

Log in to your account and learn more about what parental controls they offer.

Home Internet

How lovely would it be to simply shut off the internet?

You can do that!

Log in to your home internet service provider account. Explore the features such as timing (length of time, access by times of day/night), auto shut off, and activity monitoring.

State the rules of your home and enforce those rules by using the tools available to you. Your kids will thank you (maybe only silently, but they do need boundaries.) My daughter has told me on multiple occasions that she's grateful that I'm a strict parent.

I'll get into greater detail about how we simply shut it all down – the phone, the devices, the internet, all of it. (Check out *Chapter 7 / Create a Healthier Relationship with Tech.*)

Game Systems

Video games can be played on PCs, consoles (like Sony PlayStation, Microsoft Xbox, and the Nintendo Switch), mobile devices, in arcades, and via virtual reality (VR). There's too much variation to go into detail here.

Like any new-to-you technology, you'll want to use a combination of the manufacturer's website and well-sourced YouTube tutorials on parental control setup.

Social Media Apps

Bookmark or save any that you need from this list of major apps and platform safety centers. There are additional tools your child might be using. These are listed again in the *Resources / Tech Support & Parental Controls* section.

- Discord https://discord.com/safety-parents
- Meta (Facebook & Instagram) https://familycenter.meta.com
- Pinterest https://help.pinterest.com/en/article/teen-safety-options
- Reddit https://support.reddithelp.com/hc/en-us/articles/15484574845460-Safety
- Roblox* https://www.youtube.com/watch?v=yjjVCGEkq1U

- Snapchat https://parents.snapchat.com
- TikTok https://www.tiktok.com/safety
- Twitch https://safety.twitch.tv
- X* https://help.x.com/en/safety-and-security
- YouTube https://support.google.com/youtube/answer/15255618

Roblox is technically a video game but has chat and social components. I included it because it's so popular – and dangerous.

X does not have a family or teen-focused safety center. It is not safe for children of any age.

Now, the social media platforms have done a *terrible* job of protecting our kids from danger. That said, it's not their primary mission. That's our job. But they've received tremendous pressure to up their game, and improvements continue to be made. They aren't perfect and never will be.

There is a list of basic security settings that I think should be turned on by default. Suggestions from security specialists:

1. Set accounts to private rather than public.
2. Make friends lists private.
3. Allow only friends to comment on posts.
4. Turn off location (tracking) services.
5. Turn off personalized ads or marketing.

Social media companies make their profit from showing us ads and selling our data to advertisers. Talk to your kids regularly about this, because it's what drives all of the site decisions. Ask your kids what types of ads they're seeing – it can be an indicator of their behavior. Better yet, use the social media apps on their devices, and you'll get an accurate view of their customized experience. Learn how to adjust ad settings, report ads, and more.

The easiest way to get up-to-date instructions or specific information (such as features and ad blocking) is to Google it.

Why am I telling you to search for privacy settings information for social but not for the other tools I've written about?

The social media companies change their safety instructions rapidly. As they get pressured and laws change, they update their approach. I find they publish tech support information and change it at a faster rate than other types of tech. Every six months this information can be updated and often the original web page disappears! Bookmark the list of safety centers I've listed just above, but be ready to search all over again should some of your original resources vanish.

How do I learn about parental controls and privacy settings for Instagram, Discord, TikTok, Snapchat, and then some?

I have to Google it too.

Here's an example of what I type into the search bar: "parental controls TikTok" or "how to change privacy settings on Instagram." You can do the same thing. Just change the search terms to the name of the tool you're investigating and consider changing the filter in the search results to the last year or month to get the most recent information.

Engineers are continuously working on these sites and apps, so the tech support instructions may be old. It's frustrating; I regularly find tech support pages that don't match the current version.

You will find search results from the tech company itself, from news publications, and from tech journalists who wrote about recent changes. Read both internal and external coverage of the information. Learning more about these protections from a reliable outside source is a great way to get the good, bad, and ugly. The platforms focus on overly positive marketing messages and purposefully leave out important or negative details.

Remember, the technology companies have one thing in mind: profit. The information they provide you with doesn't tell the whole story. I recommend PC Mag (for software), Consumer Reports, TechCrunch, LifeHacker, The Verge, or WIRED (for all things tech). Any other publication you personally trust is a good resource to learn more than just the party line.

Third Party Services

There is a wide variety of apps that you can use to augment what I've already mentioned. I do not have any recommendations in this category because I don't think they're necessary. Some are very specialized, like those that help divorced families co-parent (brilliant idea!), but you have so many options available from the manufacturers and service providers that I think most tools in this category are extraneous. And they cost money. The previous resources I've given you are free.

If you like a feature that one of them offers, by golly, get it. But do you *need* a third-party app? No, you don't. Keep your money in your pocket.

Engage in Conversation

After all that, watching out for your kid and setting up parental controls, there's one more ongoing task: talking to your kids about their tech use. This isn't an interrogation. Ask questions with sincere curiosity to strengthen your relationship with your child. Ask them about their offline activities, friends, and interests – keeping that same energy for their online activities. They are intertwined.

> *"We have overprotected our children in the real world and underprotected them online"*
> Jonathan Haidt, The Anxious Generation

Your kids need to know they can trust you, that they can talk to you about all kinds of things they do and are interested in, and that you're a source of information. This is the day-to-day interaction that gives you the chance to monitor their social and emotional state, correct misconceptions, see changes as they happen, be available when they have questions, and guide them to make responsible choices.

Here are some conversation starters:

- What's your favorite app/game? Tell me about it.
- Do you talk to others while you're online? How did you meet them?

- What do you talk about? What do they talk about with others?
- Tell me about your favorite account, creator, or type of content. Why do you like them? (Ask to see the content.)
- What kinds of content do you post? How would you describe your current aesthetic?
- How do you feel when you use your favorite app/game (or any app that comes up in conversation)? Do you feel happy, sad, or something else?
- What do you not like about apps, games, and content?

There's nothing that can replace your love, interest, and reliability.

CHAPTER 7

Create a Healthier Relationship with Tech

It starts with us. As the adults in the room, we set expectations, enforce boundaries, practice intentionality in creating a positive family culture, and communicate with our kids. These are the central pillars to building a healthy relationship with technology.

We learned to parent by watching our own parents and community. Because we have no precedent when it comes to parenting children during this time of technology, it makes sense to feel unmoored, overwhelmed, or anxious. The danger is new, but you already have the skills necessary to do this.

When do we start this conversation?

Introduce healthy habits earlier than you think because technology impacts their lives from day one. We begin when they're toddlers by having open and ongoing discussions about boundaries, healthy relationships, empathy, consent, rejection, and personal development. If your child is well past the toddler stage, it's okay: start now. Our teaching begins with our everyday conversations and the behavior we model. They're always listening and watching us to pick up how they should behave as they grow.

The first step is to acknowledge your emotions.

Ask yourself these questions:

- Are you uncomfortable? Worried? Scared?
- Do you feel dense because you don't understand it all?
- Is it all moving too quickly?

Slow down and feel your feelings: identify them, and sit with them. All of your feelings are valid, and they need to be seen for you to move through them. Surprisingly, they lose much of their power over you when you see them and call them by name. You're turning on the light in a room filled with boogeymen. They're not as scary in the light.

Share your feelings of fear or being out of control with your children. You may shy away from them seeing you as less than all-powerful, as fallible, but it's an important life lesson. Honesty, vulnerability, emotional intelligence, and communication are all benefits that come from authentic conversations like these. Let your kids know that you're scared that they'll get hurt, that you have fears. It's okay to share. They need to hear it from you.

You aren't alone. Change is hard, technology is hard, and this is all hard. That doesn't mean you can't do it. You do hard things every single day, and you overcome them. You can do this too, I promise.

Build Healthy Life Habits

There are so many ways you can impact your children's relationship with technology. These are gifts you'll give your entire family. Let's dive in.

Eat Together as a Family

Make a rule to eat together at a central location. I recommend the dinner table, but the location is up to you. Will you do this multiple times a week or only on a specific day? That's your call. Try to set a pattern, and an expectation that you gather at certain times. The important part is this: no technology. No TVs, screens, or devices.

We learn a tremendous amount of our social and soft skills from talking to others. Your kids will learn how to look others in the eye (a big complaint today), hold a conversation, listen, and express empathy. Gathering regularly with family also allows you to bond and monitor their mental health (through their body language and speaking style). You'll be able to see when something isn't right. And your kids want to be seen by you.

The best part is that you'll also see and feel love, laughter, and joy. You'll share wins and losses and have important conversations. This is where life happens.

One-on-One Time

Your kids need individual bonding time with you. When you've got multiple kids, this is even more important. Not everyone is willing to talk in a group, or they may show up differently depending on the environment. This allows the two of you to delve into tougher or more sensitive topics that won't work in a loud or group setting.

Build Community

Kids need to learn from and develop relationships with people other than their parents. Help them bond with non-custodial adults to gain life perspective and support. Sports coaches, neighbors, godparents, close family friends, and religious leaders are great places to start. Our kids won't always come to us first with a problem or concern about what's going on in their lives. We want to make sure they have a strong support system.

Set up a Healthy Workspace

Our home is everything. The importance of our physical setup and how we interface with technology became *very* apparent during the COVID-19 lockdown.

During lockdown, everyone was scrambling to set up proper workspaces. We were unable to find desks, desk chairs, and other tools for home offices due to shipping constraints. Everything was backordered for months. We were stuck at our kitchen tables, on lumpy couches, and at low coffee tables for long periods of time, trying to accomplish some sort of productivity during a crisis.

We were scared and overwhelmed, and our bodies also started screaming at us because of our poor setups. Our kids likely experienced pain from their poor work arrangements. Maybe you've told them repeatedly to sit up from the puddle of goo they've drifted into on the floor, bed, or other situation, but it's worth mentioning again and again, just like any other life lesson you're teaching them. They need to know the why and the how of a healthy setup. They will hear your voice in their head in the future when you're not around to remind them.

Our physical relationship to the technology we use and how we use it, has a lifelong impact on our health. There are thousands of medical and behavioral studies based on what you're interested in, but I want to focus on ways you can improve your physical health in relation to your workspace (Mayo Clinic).[1]

Chair

Use an adjustable chair with a strong back that allows your legs to rest on the floor at a 90-degree angle. If you can't achieve that angle, use a footrest or riser to bring your thighs parallel to the floor. Adjust any armrests so your arms rest lightly and at a 90-degree angle to the surface of your desk or working area. Your arms and shoulders should feel relaxed.

Desk

It's key that your working surface is at the right height. Raise your table, desk, or surface higher up if you can. Remember 90-degree angles, both arms and legs. Pad the hard edge of the desk to reduce contact stress on your wrists.

The cost of standing desks has dropped considerably over the last decade. I got one for $120 and it's a fantastic option to reduce sedentary positions. They give you an energy boost, improve posture, reduce back pain, and keep your circulation moving among other benefits (Orthopaedic Hospital of Wisconsin).[2]

Keyboard
Keep your keyboard directly in front of you while you type rather than off to the side. Your body should feel relaxed without strain. Reaching too far forward or off-center will tire you out and become the source of soreness or pain. Bluetooth keyboards can help reduce an extended reach.

Mouse or Trackpad
Reduce the strain on your pointer finger by adjusting the number of clicks needed and the pressure required. Keep the mouse next to the

keyboard at an easy distance. Investigate the variety of ergonomic models available.

Monitors

We've gotten into an age of multiple monitors. The optimal setup for your health is a single monitor, directly in front of you. If you have an array, they should be 20-40 inches from your face (or an arm's length away). The top of the screen should be at or just below eye level.

Laptops

Many of us are using laptops instead of traditional desktop computers to do our work full-time. The setup for these devices requires extra care. Long-term use of a laptop on your actual lap or other soft surface isn't recommended due to the units overheating. Soft fabrics, such as clothing or bedding, block the fans and they become a fire hazard. That's why manufacturers prefer to call them notebooks: to discourage lap use.

Put your laptop on a riser, whether professionally manufactured or homemade, such as a box, and get it to eye level. Using Bluetooth keyboards and mice is a great improvement to prevent pain from long-term stationary use.

20-20-20 Rule

Optometrist Jeffrey Anshel came up with this easy-to-remember guideline in 1991 (Optometry Times).[3] It came from him seeing patients experiencing eye strain, headaches, and dry eyes, which has been dubbed Computer Vision Syndrome (CVS), then renamed Digital Eye Strain (DES).

20-20-20 stands for "every 20 minutes look at something 20 feet away for 20 seconds." It seems as if every computer professional can quote this guideline. The purpose is to get you to take regular breaks.

It's a catchy rule that has limited data to support the specific numbers, but keep it in mind (Modern Optometry).[4] Adjust the numbers to go up or down based on your needs.

Exercise Snacks

According to a study released in January 2024 by Columbia University,[5] exercising in small snack-sized amounts can offset the health impacts of prolonged sitting. Their recommended timing is five minutes of walking for every thirty minutes of sitting.

I don't know about you, but most of my classes and meetings run well over thirty minutes. This guideline isn't realistic for me, so I've adapted it and have my own "snacking style." I work from home, so about once an hour, I get up from my desk and do a small chore to get me moving. If I know I'll be sitting for a long time, I set a timer on my phone and take a walk for ten minutes every two hours. I also use a standing desk but honestly, I forget I even have it much of the time.

Sleep Hygiene

Our technology invades seemingly every part of our lives, including how much rest we get. Our sleep is precious and, as a society, we simply do not get enough. Proper rest impacts physical and mental health, school performance, emotional regulation, and healthy development. Set the tone for healthy sleep, and you'll give your kids a crucial building block on which the rest of their life is built.

How much should your kids sleep?

The answer depends on their age (American Academy of Sleep Medicine):[6]
- 6-12 years, 9-12 hours per day
- 13-18 years, 8-10 hours

Which kids are not getting enough sleep?

- 80% of female high school students (CDC)[7]
- 84% of twelfth-grade students
- 84% of Black high school students
- 37% of children (4 months – 5 years) (CDC)[8]
- 61% of Native Hawaiian and Pacific Islander children (4 months – 14 years)

Reduce Blue Light

Wavelengths of light and their colors have differing impacts on our health. Natural sunlight has the full spectrum of colors and wavelengths. It impacts our mood energy, metabolism, sleep, circadian rhythms, and hormone cycles. We are solar-powered.

Artificial light offers far less in terms of color and benefits. Technology has an abundance of blue wavelengths that boost attention, reaction times, and mood during the day but are disruptive to us at night. Our bodies are missing the cues that natural light gives us, and the health impacts can be severe.

Blue light disrupts your circadian rhythms and impacts sleep. This shift can increase blood sugar levels to a prediabetic state and decrease melatonin levels. Decreased sleep happens due to even low light at night and is linked to increased risk of depression, diabetes, and cardiovascular problems (Harvard Studies).[9]

Adjust the display settings on your computer and devices. Most have blue light blockers such as Apple's Night Shift[10] and Microsoft's Night Light.[11]

I have prescription eyeglasses with a blue light filter. You can also buy them just about everywhere you can get glasses, whether they're prescription or non-corrective.

Does Dark Mode help?

All of your tech and many websites offer the ability to invert the colors from light to dark. Many people prefer Dark Mode. While it does reduce eye strain, there is no definitive evidence showing that it reduces blue light disruption.

Dock All Devices

Everyone should put their devices to bed. I recommend getting a docking station that allows for multiple machines to charge via one outlet. Get the devices out of bedrooms. The separation is key for all of us to get better rest and learn how to live without technology. Many of our kids (us too!) feel anxiety when away from their tech, and that's a clear sign that they need more tech-free time in their lives. Once the phone is out of their hands, 74% of teens feel happy and 72% feel peaceful (Pew Internet Research).[12]

Put the docking station in a central or controlled location. The family living room or kitchen are good places. If your kids are sneaking their devices when they're supposed to be docked, you may need to put it in a controlled or locked room. If you can keep it out of your bedroom, that would be optimal. You'll be just as tempted to use your devices at night. Remember, we're modeling healthy behavior, and you need good sleep (and a break from your tech) too.

Nightly docking also allows you to monitor their devices. (See *Chapter 6 / Monitor Activity*). While you won't be digging through their activity every day, it's a consistent and longer span of time where you can see what they've been up to, you can move slowly and methodically, take notes on what you find, and create a plan for a calm parenting conversation that can happen the next day.

But I need my phone to set an alarm!

This is the first thing adults say when I bring up docking devices outside of the bedroom. Young people say it too, but it's lower on their objection list. My recommendation is to get a travel alarm clock, not a fancy high-tech phone dock with all the bells and whistles. We're trying to reduce tech usage. Travel alarm clocks are battery-operated, small, and low-tech. They're harder to find than you'd think, but your local drug store will have them, or you can order one online. They average about $15.

Power Down Hour

Does your brain race when you try to go to sleep at night?

When driving a car on the freeway, you don't just take the key out of the ignition to shut it down. You need to downshift your engine, get on a quieter road, then park and turn it off. Your mind needs to off-ramp from the day.

Create a bedtime routine to train the body and mind that it's time to sleep. You need this, and so do your kids. Healthy sleep doesn't happen in today's world by accident. It requires conscious effort, and below are some great ways to set yourself up for a peaceful night's rest.

Tips for getting good sleep:

- Dock and/or turn off all tech (don't keep TVs in bedrooms)
- Dim the lights
- Make tea or another warm drink
- Play soothing music
- Take a bath or shower
- Meditate or pray
- Read a physical book
- Keep bedrooms minimal in distraction and design (if possible) to encourage relaxation

Decrease Tech in Your Life

Turn Off Notifications

Would you like to receive 237 notifications each day?

Our teenagers are inundated with distractions all day long. It's no wonder they're struggling to focus in school: a quarter of those notifications arrive during school hours (CNN Health).[13]

When I don't dock my phone outside of my bedroom, I wake up and check my phone first thing. I'm looking for the time, important communications, and the weather. Often, I see a negative notification from a news app, and it sets a terrible tone for the day.

One day, I walked into a meeting in a yucky mood. I'd just seen a notification from a major news service about a horrible incident. I mentioned it to my client, and that it had an immediate impact on my mood. I realized at that moment, I could simply turn them off, and that's exactly what I did.

The experience has been liberating. I pick up my phone often enough to see the badges (the numbered bubbles on the apps) to know if I need to check something. When I realized I have control over notifications and how they impact my mental health, it was a game-changer (Instructions via Popular Science).[14]

Start by adjusting your own devices first, then turn your attention to those of your kids. Fine-tune your notifications, sounds, ringtones, vibrations (haptics), available times, and anything else that occurs to you. Dig through your settings and learn more about how to control your experience. This will benefit you and your health and will help get you more comfortable in these back areas.

My personal favorite is called Do Not Disturb for iOS. I feel downright peaceful when I turn this on. Both Google and Apple offer focus settings that allow you to flip a switch on and off to hibernate or quiet your phone in a variety of ways. Search "set up Google or Apple Focus" to learn how to adjust to your liking.

Go Gray

Our phones and their candy-colored apps are designed to be practically irresistible to our brains. Removing color or using grayscale takes away positive reinforcements and dampens the urge to keep opening your phone to shop, play games, or be social.

It's not easy to use devices stripped of color but it's a fascinating experiment and you'll notice a difference in your desire to pick it up almost immediately. Search the web for "turn on grayscale" and your device model to try it.

Set Timers

I love a timer. I use them to get me started on difficult tasks and to keep myself from slipping down a rabbit hole. When Joy struggled with transitions, such as leaving a fun place like the park or getting ready for bed, we'd set a timer. "We'll be leaving in fifteen minutes" was a phrase that saved us from many meltdowns.

Tech Tip: You can set app limits in Screen Time (iOS) or Digital Wellbeing and Parental Controls → App Timers (Android). It will give you a countdown when it's about to expire. Adults can extend use for an extra one or fifteen minutes or ignore it for the day (bypass), but minors don't have that option.

Mindful Media Consumption

Practice critical thinking with your family. Pause the movie or show and talk to your child about what's happening. Ask questions, practice media literacy.

Know the difference between passive and active use of technology. Examples of passive consumption are watching TV, movies, or videos online. Active tech use can be building or creating with technology. They are not the same. Actively using tech can be very positive and great for brain growth. Set time limits for passive and encourage active use.

Delete the App

If notifications bug you, you can go one step further. Delete the app altogether.

I found that the same news app that gave me terrible news first thing in the morning had become more and more desperate with its wording in notifications. Breaking news had devolved into useless information of the attention-seeking variety. One day, I'd had enough and deleted it from my phone. I can always get it back if I want to.

When you find that your kids are arguing with friends in-app, engaging in unhealthy behavior, or spending too much time on an app, this is a great solution. You get to decide if it's temporary or permanent.

As the parent, I recommend you make sure that only you can download apps onto their devices. This forces a discussion about what apps are appropriate and whether their recent behavior has been acceptable. It also enables you to ask questions about why they want it. Read more about it in the app store before downloading.

You decide:
Access - Devices - Timeframe

Turn it Off

There are times when the internet is too much, I'm overwhelmed, and I need a real break. This is when I turn my phone off. All the way down! And then go do something else.

Do you feel a little panicky thinking about trying that?

That means you could use this tip for yourself. Practice for twenty minutes at a time and work up to an hour. Weekends are a fantastic time to unplug.

When I was little, my dad would walk through the living room and say, "The TV is tired" and turn it off. I was disappointed but I would go find something else to do. Try it. If your kids are bored (or you are bored), that's okay. Let yourself be bored. Boredom is great for creativity.

You control your tech. It doesn't control you.

Digital Detox

How realistic is a digital detox?

I don't know how long I'd be able to do it myself. I did find an approachable way to try it: the 5:2 Digital Diet (It's Time to Log Off).[15] 5:2 stands for 5 days with tech, 2 days without. The off days are most easily done on the weekend, when many of us have less work to think about.

Here are ways to detox tech from your life during those two days:

- Use an analog alarm clock.
- Use a physical camera. Kids *love* cameras! Whether it's a film camera, a disposable, or a DSLR, let them play.
- Move more. Get out of your head and into your body.
- Get into nature. You're less likely to want tech when surrounded by nature.
- Support each other. Let your friends know what you're up to. Ask for their support or participation.
- Embrace the boredom. With time, creativity is bound to follow that boredom.

Storytime: Gaggle is Tech-Free

I have a group of girlfriends (The Gaggle) with an unusual code of conduct. We knew we had to be mindful of our media footprint. Our group included a well-known nonprofit executive, a city council member, an elementary school teacher, and me (a public figure and school teacher). Our need to not splash our fun times in the public eye was adopted by the rest of the ladies and became part of our group culture.

Early in the friendship, we made a decision not to take photos or post them to social media. Sure, we take a few pictures here and there but most of the time we don't use our devices. Upon arrival, everyone sets their phones and keys down in a central location, then we cackle, eat, drink, and are present with each other. I kind of forgot about it until my family and I were arriving home from a Gaggle pool party.

My former partner said, "That was really nice. No one used their phones."

"Of course we didn't," I replied. "We were by the pool." But he pressed the point.

"No, it wasn't that. I kept looking around, and none of your friends nor their spouses were on their phones the whole time."

I remembered our agreement and I was proud of the Gaggle.

Now, over a decade later, most of the Gaggle have kids. I know those kids will remember their moms talking and having a great time, rather than us on our phones. The downside is that we have few pictures and videos of all of us. I'll take the memories.

Get Out in Nature

Many of our kids simply can't be separated from their devices. I've taught high school and college as well as adults; I see people of all ages who struggle to unplug. But it's heartbreaking to witness a child who doesn't know how to engage in the natural world. They don't know how to play.

I took some of my Sonoma State University students out of the classroom on a tech-free campus nature walk (a mindful walk called forest bathing), and I saw something surprising. My students were in their early twenties and looking at their campus with wide eyes. They were climbing trees, having deep conversations with classmates about mushrooms they'd spotted, and meditating while watching a family of ducks. I'd never seen them so peaceful and calm. Their expressions reminded me of color-blind people seeing a rainbow for the first time. I'm not exaggerating. It seemed as if they'd never really noticed their campus before, which is known for its beauty. It was one of the most fulfilling experiences I've had as a teacher.

> *"The phone is an experience blocker."*
> Jonathan Haidt

If you teach your children to leave their devices behind, to be in nature, to live life connected to the elements around them, you will have a happier and healthier child. Our kids spend so much time indoors that it's seriously impacting their eyesight, bone development, and immune systems (World Economic Forum).[16] Nature, also known as greenspace, counteracts the impacts of modern society and technology (National Library of Medicine).[17] When I made this discovery for myself, I started calling it the Antidote to

Technology,[18] and it's a personal mission of mine to share it with anyone who will listen.

Tips for connecting with nature:

- Get outside, whether it's out in front of your house, a pocket park, or a forest. Just go.
- Have them leave their devices in the car or the house. Keep yours on you for safety reasons, but don't use it unless necessary.
- Teach your kids how to be in nature, how to sit quietly. Start with listening for sounds in nature, such as birds or the wind, rather than focusing on silence (this can make people nervous). They **need** to practice this.
- Take your shoes off and place your feet directly on the ground or in saltwater. This is called earthing or grounding and has tremendous benefits for your health. Learn more about earthing.[19]

Be Present

If you've had an absent parent, you might feel this in your gut. But I'm talking about parents who are "there but not there." When surveyed about tech usage, the biggest complaint from the children's perspective is a distracted parent. One particular study shows that 46% of teens say their parents are distracted by their phones when having conversations. But parents are not able to see their personal behavior clearly (Pew Research Center).[20]

Anytime I noticed I was on my phone when I could be present with my growing daughter, I would tuck it away and literally remind myself to be in the moment. When Joy was around seven or eight, I noticed I didn't have as many photos and videos of her as I did when she was smaller, and I had to think about why. As she grew aware of my phone, I made a more conscious effort to look her in the eye and show her I was fully there with her. She would do something adorable, and I would say to myself, "Take a mental picture" because I didn't want to break the spell and ruin it by grabbing my phone. You know what would happen if I did. I'd take the picture, then get distracted by notifications and the like. A rabbit hole for my attention.

Be there

Check back into your body. Be present for your children. Yes, I know you're stressed, tired, and overwhelmed. It's tempting to give them tech so you can get a break, but at some point, we have to acknowledge how much damage they are experiencing as a result of the convenience. Many of the things we talk about hurting our kids start with our choices. The behavior we model is the behavior they take into their adulthood.

We need to teach them how to be present as well. I like to ask my teenage students how they feel when using an app or following a particular account. I ask them to check in with their bodies.

They slowly start to describe their feelings, and it becomes easier as they continue talking. It doesn't matter what the feelings are; I'm teaching them to listen to themselves.

We are not encouraged to listen to what our bodies and minds are telling us about our experiences or surroundings. They need you to guide them in this process.

Some of the questions I ask:

- How do you feel when your phone isn't in your hand?
- How do you feel when you're using _____ app?
- How do you feel when you're looking at content by (fill in celebrity or influencer name)?
- Did you know that if it doesn't feel good, you don't have to do that?

- What else would you like to do with your time that you don't feel you have time for?

Your kids want and need your love, attention, and guidance. They need you more than any technology we can buy. They crave the real, but they don't know how to access it much of the time. They need your help.

Saying Goodbye but I'm Not Gone

We have covered so much ground together. Here's what I want you to do now: pick three things to do or work on.

Just three.

Whether it's docking phones at night, adjusting parental controls, printing the critical thinking handout and putting it on the fridge, or looking up a tutorial video on YouTube, just pick the things you can start doing now.

Then share *one thing* you've learned with a friend.

We're parents and caregivers. We take care of each other, and we take care of each other's kids. We're a team. There isn't one central location to learn everything we need to know about the tech our kids are exposed to (or anything related to raising children). We're never done learning so we need to do what loving adults have always done: we share what we know with others. We will make it through this together.

Resources

Join My Parents Group on Patreon Featuring Information on Platform Changes, Fresh Tech Support & Resources, Community Chats, and More
https://www.patreon.com/kregobiz

Book Recommendations
- "It's Complicated: The Social Lives of Networked Teens" by danah boyd

- "Earthing" by Clinton Ober, Dr Stephen T Sinatra M.D., Martin Zucker, James L Oschman Ph.D.
- "The Power of Now" Eckhart Tolle
- "The Sleep Revolution" by Arianna Huffington

Downloads

https://kerryregoconsulting.com/KKSDownloads

- Inappropriate Content | Safety Action Steps
- Critical Thinking Worksheet
- Build Critical Thinking Skills | Safety Action Steps
- Online Safety | Do's & Don'ts
- How to Protect Your Child | Safety Action Steps
- Report Abuse
- Signs Your Child is Being Bullied
- Family Safety Plan

- Increase Your Control | Safety Action Steps
- Create a Healthier Relationship with Technology

Law Enforcement & Legal Assistance

- FBI https://www.fbi.gov/how-we-can-help-you/scams-and-safety
- Social Media Victims Law Center https://socialmediavictims.org
- U.S. Department of Justice, Project Safe Childhood https://www.justice.gov/psc/national-strategy-child-exploitation-prevention-and-interdiction

Learn About Tech

- Apple classes https://www.apple.com/today
- Android https://www.android.com
- Bark (tech tips for parents) https://www.youtube.com/@BarkTechnologies
- ConnectSafely https://www.connectsafely.org
- Emoji Decoded | Drug Enforcement Administration https://www.deadiversion.usdoj.gov/mtgs/chem_industry/documents/Emoji_Drug_Code_Decoded.pdf
- GCF Global https://edu.gcfglobal.org
- LinkedIn Learning https://www.linkedin.com/learning-login/go

- Chat and Text Abbreviations and Acronyms https://messente.com/blog/text-abbreviations
- Social Media Glossary https://later.com/social-media-glossary

Regain Control

- Family Contracts https://connectsafely.org/contracts
- Family Media Plan, American Academy of Pediatrics https://www.healthychildren.org/English/fmp/Pages/MediaPlan.aspx
- Well Beings Bay Area Mental Health Resource Toolkit | WETA & KQED https://drive.google.com/file/d/1cOFSDczi5-tCP0LaT0EB69SMsdDtFPTV/view

Tech Support & Parental Controls
Computers & Devices

- Microsoft Parental Controls
  https://www.microsoft.com/en-us/microsoft-365/family-safety
- Apple Parental Controls (Computer)
  https://apple.co/3JPyiTT
- Apple Parental Controls (Devices)
  https://support.apple.com/en-us/105121
- Google Chrome Parental Controls (Computer)
  https://support.google.com/chromebook/answer/7680868?hl=en
- Google Family Link (Devices)
  https://families.google/familylink

- Google, Monitor Activity
  https://myactivity.google.com/myactivity

Social Media

- Discord https://discord.com/safety-parents
- Google, Remove Info from Search
  https://support.google.com/websearch/answer/9673730
- Meta (Facebook & Instagram)
  https://familycenter.meta.com
- Pinterest
  https://help.pinterest.com/en/article/teen-safety-options
- Reddit
  https://support.reddithelp.com/hc/en-us/articles/15484574845460-Safety
- Roblox Parental Controls
  https://www.youtube.com/watch?v=yjjVCGEkq1U
- Snapchat https://parents.snapchat.com
- TikTok https://www.tiktok.com/safety

- Twitch https://safety.twitch.tv
- X* https://help.x.com/en/safety-and-security
- YouTube
    - Family Center
      https://support.google.com/youtube/answer/15255618
    - Check YouTube watching history
      https://support.google.com/youtube/answer/95725
    - Turn on/off YouTube Restricted Mode
      https://support.google.com/youtube/answer/174084
    - Report YouTube content
      https://support.google.com/youtube/answer/2802027

Bibliography

Chapter 2 | Everything Feels Out of Control

1. International Certification of Digital Literacy (ICDL)
   https://icdl.org
2. New CDC data illuminate youth mental health threats during the COVID-19 pandemic | Centers for Disease Control (CDC)
   https://archive.cdc.gov/#/details?url=https://www.cdc.gov/media/releases/2022/p0331-youth-mental-health-covid-19.html
3. Moral Panic, Scott A. Bonn Ph.D. | Psychology Today
   https://www.psychologytoday.com/us/blog/wicked-deeds/201507/moral-panic-who-benefits-public-fear
4. Newspaper bad phone good boomer bad | Reddit
   https://www.reddit.com/r/im14andthisisdeep/comments/dfinq0/newspaper_bad_phone_good_boomer_bad
5. Children and Adolescents and Digital Media | American Academy of Pediatrics
   https://publications.aap.org/pediatrics/article/138/5/e20162593/60349/Children-and-Adolescents-and-Digital-Media
6. Techno-legal Solutionism: Regulating Children's Online Safety in the United States, María P. Angel

and danah boyd | ACM Digital Library https://dl.acm.org/doi/10.1145/3614407.3643705?ref=made-not-found-by-danah-boyd.ghost.io
7. KOSA Isn't Designed To Help Kids, danah boyd | LinkedIn https://www.linkedin.com/pulse/kosa-isnt-designed-help-kids-danah-boyd-bkpbc
8. Children's Online Privacy Protection Act (COPPA) | Federal Trade Commission https://www.ftc.gov/legal-library/browse/rules/childrens-online-privacy-protection-rule-coppa
9. US Surgeon General says 13 is too young to join social media | CNN https://www.cnn.com/videos/business/2023/01/29/vivek-murthy-social-media-13-too-young-brown-nr-sot-vpx-contd.cnn

Chapter 3 | The Danger Zones

1. FTC distributes $5.6 million in refunds to Ring customers from privacy settlement, Zo Ahmed | Techspot https://www.techspot.com/news/102774-ftc-distributes-56-million-refunds-ring-customers-privacy.html
2. What's Your Data Really Worth? Ben Wolford | Proton https://proton.me/blog/what-is-your-data-worth

3. Facebook Admits to U.S. Senator - We Lied About Our Teenage Spy App, Zak Doffman | Forbes https://www.forbes.com/sites/zakdoffman/2019/03/02/facebook-caught-lying-about-spying-on-teenage-users-data-yet-again
4. Google to pay California $93 million for allegedly lying to users about location data practices, Suzanne Smalley | The Record https://therecord.media/google-settles-for-lying-geolocation
5. Gonzalez v. Google (2003), Deborah Fisher | Free Speech Center https://firstamendment.mtsu.edu/article/gonzalez-v-google
6. What Are the Most Played Songs in Radio History? Jordan Potter | Far Out https://faroutmagazine.co.uk/most-played-songs-on-radio
7. Android vs. iOS: Security Comparison 2025, Daniel Markuson | NordVPN https://nordvpn.com/blog/ios-vs-android-security
8. Data Breach Class Action Claims Temu's In-App Web Browser Secretly Tracks Users, Kelly Mehorter | ClassAction https://www.classaction.org/news/data-breach-class-action-claims-temus-in-app-web-browser-secretly-tracks-users

9. Get the U.S.'s Top Apps of 2024 | Apple App Store
   https://apps.apple.com/story/id1778539371
10. Most Popular Apps, David Curry | Business of Apps
    https://www.businessofapps.com/data/most-popular-apps
11. Bluetooth Security Risks to Know (and How to Avoid Them), Clare Stouffer | Norton
    https://us.norton.com/blog/mobile/bluetooth-security
12. U.S. School District Spied on Students Through Webcams, Daniel Nasaw | The Guardian
    https://www.theguardian.com/world/2010/feb/19/schools-spied-on-students-webcams
13. How the USA Patriot Act Expands Law Enforcement "Sneak and Peek" Warrants | ACLU
    https://www.aclu.org/issues/national-security/privacy-and-surveillance/surveillance-under-patriot-act
14. Mark Zuckerberg Puts Tape Over His Webcam | ABC News
    https://abcnews.go.com/Technology/mark-zuckerberg-puts-tape-webcam/story?id=40040340
15. Deepfake | Oxford Dictionary
    https://www.oxfordlearnersdictionaries.com/definition/english/deepfake
16. High School Student Allegedly Used Real Photos to Create Pornographic 'Deepfakes' of Female Classmates, Marisa Sarnoff | Law & Crime

https://lawandcrime.com/high-profile/high-school-student-allegedly-used-real-photos-to-create-pornographic-deepfakes-of-female-classmates

17. Beverly Hills Middle School Rocked by AI-Generated Nude Images of Students, Jon Healey | Los Angeles Times https://www.latimes.com/california/story/2024-02-26/beverly-hills-middle-school-is-the-latest-to-be-rocked-by-deepfake-scandal

18. Mark Zuckerberg Concealed His Kids' Faces on Instagram. Should You? Samantha Murphy Kelly | CNN https://www.cnn.com/2023/07/09/tech/mark-zuckerberg-emoji-kids-faces/index.html

19. 5 Biggest Risks of Sharing Your DNA with Consumer Genetic-Testing Companies, Eric Rosenbaum | CNBC https://www.cnbc.com/2018/06/16/5-biggest-risks-of-sharing-dna-with-consumer-genetic-testing-companies.html

20. The Privacy Problems of Direct-to-Consumer Genetic Testing, Catherine Roberts | Consumer Reports https://www.consumerreports.org/health/dna-test-kits/privacy-and-direct-to-consumer-genetic-testing-dna-test-kits-a1187212155

21. Hacker Leaks Millions of New 23andMe Genetic Data Profiles, Lawrence Abrams | Bleeping

Computer https://www.bleepingcomputer.com/news/security/hacker-leaks-millions-of-new-23andme-genetic-data-profiles

22. 23andMe is filing for bankruptcy. Here's what it means for your genetic data, Joe Hernandez | NPR https://www.npr.org/2025/03/24/nx-s1-5338622/23andme-bankruptcy-genetic-data-privacy

23. Attorney General Bonta Urgently Issues Consumer Alert for 23andMe Customers | California Department of Justice https://oag.ca.gov/news/press-releases/attorney-general-bonta-urgently-issues-consumer-alert-23andme-customers

24. Google to Delete Billions of Browser Records to Settle 'Incognito' Lawsuit, Catherine Thorbecke | CNN https://www.cnn.com/2024/04/01/tech/google-to-delete-data-records-to-settle-incognito-lawsuit

25. Incognito Browser: What It Really Means | Mozilla https://www.mozilla.org/en-US/firefox/browsers/incognito-browser

26. Guidance on the Protection of Personally Identifiable Information (PII) | U.S. Department of Labor https://www.dol.gov/general/ppii

27. The Unknown Danger of Child Identity Theft, Steve Weisman | Forbes

https://www.forbes.com/sites/steveweisman/2024/09/13/the-unknown-danger-of-child-identity-theft
28. Does WiFi Affect the Brain? Dr. Sanchari Sinha Dutta, Ph.D. | News-Medical https://www.news-medical.net/health/Does-WiFi-Affect-the-Brain.aspx

Chapter 4 | Staying Safe Online: Safety Action Steps

1. CyberTipline, National Center for Missing & Exploited Children (NCMEC) https://report.cybertip.org
2. News Literacy in America: A Survey of Teen Information Attitudes, Habits & Skills (2024) News Literacy Project https://newslit.org/news-literacy-in-america
3. Aging in an Era of Fake News, Nadia M. Brashier & Daniel L. Schacter | National Library of Medicine https://pmc.ncbi.nlm.nih.gov/articles/PMC7505057
4. Why Do Americans Share So Much Fake News? One Reason is They Aren't Paying Attention, Denise-Marie Ordway | Nieman Lab https://www.niemanlab.org/2021/03/why-do-americans-share-so-much-fake-news-one-big-reason-is-they-arent-paying-attention-new-research-suggests
5. News Literacy Project https://newslit.org

6. Senate Intel Committee Releases Bipartisan Report on Russia's Use of Social Media |U.S. Senate Select Committee on Intelligence https://www.intelligence.senate.gov/press/senate-intel-committee-releases-bipartisan-report-russia%E2%80%99s-use-social-media
7. Did Daisy the Dog Rescue Hundreds of 9/11 Survivors? Barbara Mikkelson | Snopes https://www.snopes.com/fact-check/daisy-dog-rescued-911-survivors
8. Snopes https://www.snopes.com
9. FactCheck https://www.factcheck.org
10. How to Avoid "Inspiration Porn", Andrew Pulrang | Forbes https://www.forbes.com/sites/andrewpulrang/2019/11/29/how-to-avoid-inspiration-porn/?sh=7be289f75b3d
11. What Are the Risks of Clicking on Malicious Links? Jasdev Dhaliwal | McAfee https://www.mcafee.com/blogs/internet-security/what-are-the-risks-of-clicking-on-malicious-links
12. How to Recognize and Report Spam Text Messages | FTC https://consumer.ftc.gov/articles/how-recognize-and-report-spam-text-messages

13. Defining Sexual Exploitation and Abuse and Sexual Harassment | UN Refugee Agency
https://www.unhcr.org/us/what-we-do/how-we-work/tackling-sexual-exploitation-abuse-and-harassment/what-sexual-exploitation
14. Online Grooming Guide | Internet Matters
https://www.internetmatters.org/resources/online-grooming-guide-what-parents-need-to-know
15. Child Sexual Abuse Material (CSAM) | U.S. Department of Justice
https://drive.google.com/file/d/1LWWhzzmTTks6Hjr5pfzrwX9M8RiTGhDF/view
16. In Their Own Words Cybersurvey Report 2019 | Internet Matters & Youthworks
https://www.internetmatters.org/about-us/in-their-own-words-2019-cybersurvey-research-report
17. Children and Parents: Media Use and Attitudes Report 2020-21 | Ofcom
https://drive.google.com/file/d/1bpDyolmUftKz42TQqUvyk_ts7qVZbXAH/view
18. Technology Report | U.S. Department of Justice
https://drive.google.com/file/d/1ohy6J6UnBzrFTme-aGhnbq6N8PRdyjxk/view
19. How Do High-Risk Youth Use the Internet? Characteristics and Implications for Prevention, Melissa Wells & Kimberly J. Mitchell | Sage Journals

https://journals.sagepub.com/doi/abs/10.1177/1077559507312962

20. State, Territory, and Tribal Resources | Child Welfare Information Gateway https://www.childwelfare.gov/resources/states-territories-tribes

21. CyberTipline | National Center for Missing & Exploited Children (NCMEC) https://report.cybertip.org

22. Take It Down | National Center for Missing & Exploited Children (NCMEC) https://takeitdown.ncmec.org

23. Resources For Survivors of Child Sexual Abuse | Enough Abuse https://enoughabuse.org/get-help/survivor-support

24. The Porn Conversation https://thepconversation.org

25. Revenge Porn: Laws & Penalties, Janet Portman updated by Rebecca Pirius | Criminal Defense Lawyer by NOLO https://www.criminaldefenselawyer.com/resources/revenge-porn-laws-penalties.htm

Chapter 5 | Bullying

1. Boys' and Girls' Relational and Physical Aggression in Nine Countries | National Library of Medicine https://pmc.ncbi.nlm.nih.gov/articles/PMC3736589

2. The Concerns and Challenges of Being a U.S. Teen: What the Data Show | Pew Research Center
   https://www.pewresearch.org/short-reads/2019/02/26/the-concerns-and-challenges-of-being-a-u-s-teen-what-the-data-show
3. Black teens more likely than those who are Hispanic or White to say they have been cyberbullied because of their race or ethnicity | Pew Research Center
   https://www.pewresearch.org/internet/2022/12/15/teens-and-cyberbullying-2022/pi_2022-12-13_teens-cyberbullying_0-05-png
4. Black or Hispanic teens are far more likely than White teens to say online harassment and bullying are a major problem for people their age | Pew Research Center
   https://www.pewresearch.org/internet/2022/12/15/teens-and-cyberbullying-2022/pi_2022-12-13_teens-cyberbullying_0-06-png
5. Youth Risk Behavior Survey, United States, 2023 | Centers for Disease Control (CDC)
   https://www.cdc.gov/mmwr/volumes/73/su/su7304a6.htm
6. Chart, Based on The State of Bullying in Schools | Education Week
   https://www.edweek.org/leadership/the-state-of-bullying-in-schools-in-charts/2023/08

7. Teens and Cyberbullying 2022, Emily A. Vogels | Pew Research Center https://www.pewresearch.org/internet/2022/12/15/teens-and-cyberbullying-2022
8. 9 Facts About Bullying in the U.S., Katherine Schaeffer| Pew Research Center
9. https://www.pewresearch.org/short-reads/2023/11/17/9-facts-about-bullying-in-the-us
10. U.S. Department of Health & Human Services, https://www.stopbullying.gov
11. Get Help Now | U.S. Department of Health & Human Services https://www.stopbullying.gov/resources/get-help-now
12. Stomp Out Bullying, Help Chat Crisis Line https://www.stompoutbullying.org/helpchat
13. National Center for Missing & Exploited Children (NCMEC) https://www.missingkids.org
14. Prevent Cyberbullying | U.S. Department of Health & Human Services https://www.stopbullying.gov/cyberbullying/prevention
15. Federal Laws | U.S. Department of Health & Human Services https://www.stopbullying.gov/resources/laws/federal

Chapter 6 | How to Regain Control

1. GCF Global Learning https://edu.gcfglobal.org
2. Connect Safely https://www.connectsafely.org
3. Apple In-Store Classes
   https://www.apple.com/today
4. Android/Google Device Options
   https://www.android.com
5. LinkedIn Learning via Your Local Library
   https://www.linkedin.com/learning-login/go
6. ConnectSafely Technology Contracts
   https://connectsafely.org/contracts
7. A Window Into Young Children's Online Worlds | Ofcom https://www.ofcom.org.uk/media-use-and-attitudes/media-habits-children/a-window-into-young-childrens-online-worlds
8. Two Factor Authentication Keeps Your Assets Safe | Kerry Rego Consulting https://kerryregoconsulting.com/2022/06/24/twofactorauthentication
9. Top 250+ Text Abbreviations and Acronyms | Messente https://messente.com/blog/text-abbreviations
10. Social Media Glossary | Later
    https://later.com/social-media-glossary
11. Parents 'Don't Use' Parental Controls on Facebook and Instagram, Dan Milmo | The Guardian https://www.theguardian.com/technology/2024/sep/1

2/parental-controls-facebook-instagram-meta-nick-clegg

12. YouTube, I Bought My Kid a New Phone
    https://www.youtube.com/watch?v=G_TyEaLbtzw

Chapter 7 | Create a Healthier Relationship with Tech

1. Office Ergonomics | Mayo Clinic
   https://www.mayoclinic.org/healthy-lifestyle/adult-health/in-depth/office-ergonomics/art-20046169
2. Standing Desk Ergonomics: 7 Benefits of Standing at Work | Orthopaedic Hospital of Wisconsin
   https://www.ohow.com/2021/02/08/standing-desk-ergonomics-7-benefits-of-standing-at-work
3. Deconstructing the 20-20-20 Rule for Digital Eye Strain, Brian Chou | Optometry Times
   https://www.optometrytimes.com/view/deconstructing-20-20-20-rule-digital-eye-strain
4. Myth-Busting the 20-20-20 Rule, Andrew D. Pucker | Modern Optometry
   https://modernod.com/articles/2023-july-aug/myth-busting-the-202020-rule
5. Rx for Prolonged Sitting: A 5 Minute Stroll Every Half Hour | Columbia University
   https://www.cuimc.columbia.edu/news/rx-prolonged-sitting-five-minute-stroll-every-half-hour

6. Recommended Amount of Sleep for Pediatric Populations: A Consensus Statement of the American Academy of Sleep Medicine | Journal of Clinical Sleep Medicine https://jcsm.aasm.org/doi/10.5664/jcsm.5866
7. FastStats: Sleep in High School Students | Centers for Disease Control (CDC) https://www.cdc.gov/sleep/data-research/facts-stats/high-school-students-sleep-facts-and-stats.html
8. FastStats: Sleep in Children | Centers for Disease Control (CDC) https://www.cdc.gov/sleep/data-research/facts-stats/children-sleep-facts-and-stats.html
9. Blue Light Has a Dark Side | Harvard Medical School https://www.health.harvard.edu/staying-healthy/blue-light-has-a-dark-side
10. Use Night Shift on Your Mac | Apple https://support.apple.com/en-us/102191
11. Change Display Brightness and Color in Windows | Microsoft https://support.microsoft.com/en-us/windows/change-display-brightness-and-color-in-windows-3f67a2f2-5c65-ceca-778b-5858fc007041
12. How Teens and Parents Approach Screen Time, Monica Anderson, Michelle Faverio, and Eugenie Park | Pew Research Center

https://www.pewresearch.org/internet/2024/03/11/how-teens-and-parents-approach-screen-time

13. Teens Are Exhausted by Phone Notifications But Don't Know How to Quit, Kristen Rogers | CNN Health
14. https://www.cnn.com/2023/09/26/health/teen-hundreds-of-phone-notifications-report-wellness
15. The Single Best Way to Make Your Smartphone Less Stressful, Eleanor Cummins | Popular Science
https://www.popsci.com/fight-smartphone-stress-turn-off-notification-badges
16. 5:2 Digital Diet | It's Time to Log Off
https://www.itstimetologoff.com/5-2-digital-diet
17. Children aren't getting enough sunlight, and it's affecting their sight, Vybarr Cregan-Reid | World Economic Forum
https://www.weforum.org/stories/2018/11/modern-life-offers-children-almost-everything-they-need-except-daylight
18. The Health Benefits of the Great Outdoors: A Systematic Review and Meta-Analysis of Greenspace Exposure and Health Outcomes, Caoimhe Twohig-Bennett and Andy Jones | National Library of Medicine
https://pmc.ncbi.nlm.nih.gov/articles/PMC6562165

19. The Antidote to Technology | Kerry Rego Consulting
    https://kerryregoconsulting.com/2012/04/29/the-antidote-to-technology
20. What is Grounding | Kerry Rego Consulting
    https://kerryregoconsulting.com/2025/02/16/what-is-grounding
21. How Teens and Parents Approach Screen Time, Monica Anderson, Michelle Faverio, and Eugenie Park | Pew Research Center
    https://www.pewresearch.org/internet/2024/03/11/how-teens-and-parents-approach-screen-time

ABOUT THE AUTHOR

Based in the San Francisco Bay Area, Kerry Rego is a social media consultant, trainer, author, and speaker working with individuals, businesses, government, and nonprofits since 2006. She is passionate about creating strategies to achieve her clients' marketing objectives, implementing and fixing tools, training staff, and providing data-driven resources.

Specializing in social media marketing, Kerry is associate faculty at Santa Rosa Junior College, a subject matter expert for California Community Colleges, and a former lecturer at CSU Sonoma State University. With over 1300 hours in the classroom and 950 presentations under her belt, she has educated and entertained audiences all over the world.

Kerry is the author of four books on social media:

- *Keeping Your Kids Safe on the Internet: A Parent's Guide to Technology*
- *The Social Media Starter Kit & companion workbook*
- *What You Don't Know About Social Media CAN Hurt You: Take Control of Your Online Reputation.*

Learn more: KerryRegoConsulting.com

www.ingramcontent.com/pod-product-compliance
Lightning Source LLC
LaVergne TN
LVHW052257070426
835507LV00036B/3100